Instability Constants
of
Complex Compounds

Instability Constants
of
Complex Compounds

by
K. B. YATSIMIRSKII
and
V. P. VASIL'EV

Translated from the Russian by
D. A. PATERSON

Translation Editor:
R. H. PRINCE Ph. D.
University Demonstrator in Chemistry, University of Cambridge

PERGAMON PRESS
OXFORD · LONDON · NEW YORK · PARIS

1960

PERGAMON PRESS LTD.,
4 and 5 Fitzroy Square, London W.1.
Headington Hill Hall, Oxford.

PERGAMON PRESS INC.,
122 East 55th Street, New York 22, N.Y.
1404 New York Avenue N.W., Washington 5, D.C.
P.O. Box 47715, Los Angeles, California.

PERGAMON PRESS S.A.R.L.
24 Rue des Écoles, Paris Vᵉ.

PERGAMON PRESS G.mb.H.
Kaiserstrasse 75, Frankfurt-am-Main.

ISBN 978-1-4684-8406-9   ISBN 978-1-4684-8404-5 (eBook)
DOI   10.1008/978-1-4684-8404-5

Library of Congress Card Number, 60-10220

C O N T E N T S

THEORY

CONTENTS

Set by Patricia Birmingham

PREFACE

In recent years many research workers have turned their attention to the quantitative characterization of complex compounds and reactions of complex-formation in solution. Instability constants characterize quantitatively the equilibria in solutions of complex compounds and are extensively used by chemists of widely-varying specialities, in analytical chemistry, electrochemistry, the technology of non-ferrous and rare metals, etc., for calculations of various kinds.

Despite the wealth of numerical data, no reasonably full collection of instability constants of complex compounds has been made until now. The various individual collections of data are far from complete and in most cases omit references to the source materials. Moreover, the present state of the chemistry of complex compounds most urgently demands the complete systematization of data on instability constants and an extension of work in this field which would take advantage of the latest physico-chemical methods.

The present work contains instability constants for 1,381 complex compounds.

We have considered it convenient to preface the summary of the instability constants with an introductory section of a general theoretical character. This section deals with methods for the calculation of instability constants from experimental data, the influence of external conditions, such as temperature and ionic strength, on the stability of complexes, and the principal factors determining the stability of complex compounds in aqueous solution.

PREFACE

In compiling the summary we have used the original literature and abstracts for the most part up to 1954, and some
work published in 1955-1956.

It should be noted that, despite our thorough survey of
the literature, it is not impossible that individual works
may have escaped our notice.

All notices pointing out errors, making further suggestions
or reporting omissions from the summary will be gratefully
received.

K. B. Yatsimirskii.
V. P. Vasil'ev.

THEORY

C H A P T E R   I

FUNCTIONS DEFINING STEPWISE

COMPLEX-FORMATION IN SOLUTION

The chief thermodynamic characteristic of a complex particle* in solution is the equilibrium constant for the dissociation of the complex into its component central ion and
ligand or ligands. Physico-chemical studies to-day on the
determination of such equilibrium constants in aqueous electrolyte solutions are often carried out using solutions of
constant ionic strength. This makes it possible to use the
equilibrium concentrations of the substances taking part in
the reaction, instead of their activities, in the equation
for the constant according to the law of mass action**.

The dissociation of a complex particle MA_n in solution
may be represented by the equation:

$$MA_n \rightleftharpoons M + nA \qquad (\text{I. 1})$$

(the charges on the particles are omitted for simplicity).

---

*   In the present work the terms "complex compound" and
"complex" are used, together with the synonymous term
"complex particle". The three terms are used about
equally in the literature on complex compounds in our
country and abroad. The terms complex compound, complex particle and complex signify a particle formed by
two or more particles capable of independent existence
in solution (one of these particles is usually a metal
ion).

** See p. 59 for more details.

The equilibrium constant for the process (I. 1)

$$K_n = \frac{[M] \cdot [A]^n}{[MA_n]}$$ (I. 2)

is known as the instability constant of the complex particle MA_n.

The instability constant is the most objective characteristic of the stability of a complex particle in solution as regards dissociation, since in the physical sense of the definition it is independent of the concentration conditions (the pH of the medium, excess reagent, etc.) and of the method used in the determination.   It is related to the free energy change on complex-formation by the well-known thermodynamic relationship

$$\Delta G = RT \ln K_n$$

In actual fact, the breakdown of a complex particle is less simple in character: like the dissociation of polybasic acids, it takes place in stages:

$$MA_n \rightleftarrows MA_{n-1} + A,$$ (I. 3)

$$MA_{n-1} \rightleftarrows MA_{n-2} + A.$$ (I. 4)

The products of the stepwise dissociation of the complex MA_n, MA_{n-1}, MA_{n-2},...., MA are present in solution in varying amounts, depending on the concentration of ligand.   The equilibria between these particles are described by a series of equations of the type (I. 3).

The corresponding equilibrium constants

$$k_n = \frac{[MA_{n-1}] \cdot [A]}{[MA_n]},$$ (I.3a)

$$k_{n-1} = \frac{[MA_{n-2}] \cdot [A]}{[MA_{n-1}]}$$ (I.4a)

are usually known as the intermediate, or stepwise instability constants, to distinguish them from the total or overall instability constant given by equation (I. 2).

The intermediate constants are related to the overall constant by the simple relationship

$$K_n = k_1 \cdot k_2 \ldots k_{n-1} \cdot k_n. \qquad (I. 5)$$

The term "overall instability constant" can of course be applied not only to the complexes with the maximum co-ordination saturation, but also to all the other particles formed in solution as a result of stepwise dissociation.

Thus, for example [1], the reaction between soluble nickel compounds and ammonia in aqueous solution leads to the formation of the ions $NiNH_3^{2+}, Ni(NH_3)_2^{2+}, Ni(NH_3)_3^{2+}, Ni(NH_3)_4^{2+},$ $Ni(NH_3)_5^{2+},$ and $Ni(NH_3)_6^{2+}$. Each particle formed is present in equilibrium with the other products of stepwise dissociation, including ammonia and the nickel ion.

For the simplest complex $NiNH_3^{2+}$ the overall constant and intermediate constant are the same:

$$NiNH_3^{2+} \rightleftharpoons Ni^{2+} + NH_3, \qquad (I. a)$$

$$K_1 = k_1 = \frac{[Ni^{2+}] [NH_3]}{[NiNH_3^{2+}]}. \qquad (I. 6)$$

For the particle $Ni(NH_3)_2^{2+}$ the intermediate dissociation equilibrium is given by the equation:

$$Ni(NH_3)_2^{2+} \rightleftharpoons NiNH_3^{2+} + NH_3, \qquad (I. b)$$

and the overall dissociation equilibrium by the equation

$$Ni(NH_3)_2^{2+} \rightleftharpoons Ni^{2+} + 2NH_3. \qquad (I. c)$$

The intermediate instability constant for the complex $Ni(NH_3)_2^{2+}$ is then

$$k_2 = \frac{[Ni(NH_3)^{2+}] \cdot [NH_3]}{[Ni(NH_3)_2^{2+}]}, \qquad (I. 7)$$

and the overall instability constant is

$$K_2 = \frac{[Ni^{2+}] \cdot [NH_3]^2}{[Ni(NH_3)_2^{2+}]}. \qquad (I. 8)$$

For the particle with maximum co-ordination, $Ni(NH_3)_6^{2+}$, the intermediate and overall dissociations are given by equations (I.d) and (I.e) respectively:

$$Ni(NH_3)_6^{2+} \rightleftharpoons Ni(NH_3)_5^{2+} + NH_3, \qquad (I.\ d)$$

$$Ni(NH_3)_6^{2+} \rightleftharpoons Ni^{2+} + 6NH_3, \qquad (I.\ e)$$

and the intermediate and overall instability constants by the expressions (I.9) and (I.10) respectively.

$$k_6 = \frac{[Ni(NH_3)_5^{2+}] \cdot [NH_3]}{[Ni(NH_3)_6^{2+}]}, \qquad (I.\ 9)$$

$$K_6 = \frac{[Ni^{2+}] \cdot [NH_3]^6}{[Ni(NH_3)_6^{2+}]}. \qquad (I.10)$$

In the literature the term "stability constant", which is the reciprocal of the instability constant, is often found in addition to the term "instability constant".

$$K_{stab} = \beta = \frac{1}{K_n}, \qquad (I.11)$$

and similarly, the intermediate stability constant:

$$k_{stab} = \varkappa = \frac{1}{k_n}. \qquad (I.12)$$

It is obvious that

$$\beta_n = \varkappa_1 \varkappa_2 \ldots \varkappa_{n-1} \varkappa_n. \qquad (I.5a)$$

The term "complexity constant", proposed by Bjerrum [1], and the term "dissociation constant of the complex" are much less extensively used.

Intermediate instability constants are of the greatest importance for the understanding of the most varied processes in chemical technology (in hydrometallurgy and electrochemistry, for example) and in chemical analysis. In this connexion, the vast majority of works devoted to the study of ionic equilibria in complex-formation and published within the last decade take into consideration, in some form or other, the stepwise nature of the process of complex-formation, and in almost all cases the intermediate instability constants are calculated.

In order to find these constants it is convenient to use a number of functions which can be readily calculated from experimental data and which in turn are related by fairly simple expressions to the instability constants. It is often the case that the mathematical expression of these relation-·ships has a simpler form if the stability constants are used instead of the instability constants. We shall therefore hereinafter make use of both stability constants (particularly for expressing mathematical relationships) and instability constants.

Some of these functions defining stepwise complex-formation have been used very extensively in recent years.

A number of interesting mathematical methods have been proposed for the calculation of instability constants using these functions, but use of them is limited to the potentiometric method and methods involving ion exchange, distribution between two solvents and, to a certain extent, polarography.

At the same time, the possibility exists today of a unified approach to the study of equilibria in stepwise complex-formation and of a considerable extension of the limits of application of existing methods for treating experimental data.

In recent years the most extensive use has been made of the so-called "formation function" $\bar{n}$, proposed by Bjerrum [1]:

$$\bar{n} = \frac{c_a - [A]}{c_M} , \qquad (I.13)$$

where c_a and c_M are the total concentrations of ligand and complex-forming metal ion in solution, and [A] is the equilibrium ligand concentration.

The formation function n is the ratio of the concentration of ligand bound in the complex (or complexes) to the total concentration of complex-forming metal ion. The physical significance of the formation function is that it represents the characteristic average co-ordination number, and it may vary from zero, when no complex-formation takes place in the solution $(c_a = [A])$, to the limiting maximum value of the co-ordination number. In the absence of stepwise complex-formation and with a sufficient excess of ligand, $\bar{n}$ is equal to the co-ordination number.

The function $\bar{n}$ is related to the instability constants by the following relationship:

$$\bar{n} = \frac{\beta_1[A] + 2\beta_2[A]^2 + \ldots + n\beta_n[A]^n}{1 + \beta_1[A] + \beta_2[A]^2 + \ldots + \beta_n[A]^n}, \qquad (I.14)$$

or, in abbreviated form:

$$\bar{n} = \frac{\sum\limits_{i=1}^{i=n} i\beta_i[A]^i}{1 + \sum\limits_{i=1}^{i=n} \beta_i[A]^i}. \qquad (I.14a)$$

Leden [2], Fronaeus [3] and a number of other workers have made extensive use of the function Φ, which is the ratio of the total concentration of metal c_M to the equilibrium concentration of free metal ions:

$$\Phi = \frac{c_M}{[M]}. \qquad (I.15)$$

Yatsimirskii [4] has suggested naming this function the "complexity" or "degree of complex formation", since it defines the extent to which complex-formation has proceeded in a given system. The degree of complex-formation may vary from 1 (in the absence of complex-formation $c_M = [M]$) to very large values, depending on the instability constants and ligand concentration.

Since the total concentration of metal in solution (c_M) is the sum of the concentrations of complexes of the type MA_i, then obviously:

$$\Phi = 1 + \beta_1[A] + \beta_2[A]^2 + \ldots + \beta_n[A]^n, \qquad (I.16)$$

or, in abbreviated form:

$$\Phi = 1 + \sum\limits_{i=1}^{i=n} \beta_i[A]^i. \qquad (I.16a)$$

It is sometimes convenient to find the fraction of a given complex, defined as the ratio of the concentration of the complex (MA_m) to the total concentration of metal in solution (c_M).

$$\alpha_m = \frac{[MA_m]}{c_M}. \qquad (I.17)$$

The quantity α_m may vary between 0 (when the complex is absent) and 1 (when other complexes are absent). Strictly speaking, α_m varies between a value close to zero and a value close to unity, since when complex-formation takes place in any solution the complex-forming metal ion is present in at least two forms M_{aq} and MA_m.

The relationship between α_m and the instability constants is given by the equation:

$$\alpha_m = \frac{\beta_m [A]^m}{1 + \beta_1 [A] + \beta_2 [A]^2 + \ldots + \beta_n [A]^n} , \qquad (I.18)$$

or, on simplification:

$$\alpha_m = \frac{\beta_m}{[A]^{-m} + \beta_1 [A]^{1-m} + \ldots + \beta_n [A]^{n-m}} , \qquad (I.18a)$$

in abbreviated form:

$$\alpha_m = \frac{\beta_m}{\sum_{i=0}^{i=m} \beta_i [A]^{i-m}} . \qquad (I.18b)$$

As the concentration of ligand is increased, α_m passes through a maximum. It can be shown [5] that at the maximum point

$$\alpha_m = \frac{\sqrt{\dfrac{\varkappa_m}{\varkappa_{m+1}}}}{2 + \sqrt{\dfrac{\varkappa_m}{\varkappa_{m+1}}}} , \qquad (I.19)$$

$$[A] = \frac{1}{\sqrt{\varkappa_m \varkappa_{m+1}}} . \qquad (I.20)$$

The formation function, degree of complex-formation and fraction of a given complex are related mathematically to one another. This question has been partially dealt with in a number of works [6-8].

Differentiating equation (I.16), we have:

$$\frac{\partial \Phi}{\partial [A]} = \sum_{i=1}^{i=n} i \beta_i [A]^{i-1}. \qquad (I.21)$$

From equations (I.21) and (I.14a) it follows that:

$$\bar{n} = \frac{\frac{\partial \Phi}{\partial [A]} [A]}{\Phi} .$$

(I.22)

After simple conversion we have:

$$\bar{n} = \frac{\partial \log \Phi}{\partial \log [A]} .$$

(I.23)

The value of $\bar{n}$ may be found by graphical differentiation of Φ.   If we plot the logarithm of the degree of complex-formation against the logarithm of the equilibrium ligand concentration, the gradient of the tangent to the curve at any point gives the value of $\bar{n}$.

From equation (I.23) it follows that:

$$\log \Phi = \int \bar{n} \, d\log[A] + B.$$

(I.24)

If the formation function $(\bar{n})$ is known, the degree of complex-formation may be found by graphical integration of the curve log [A] vs. $\bar{n}$.

From equations (I.16) and (I.18) it follows that

$$\alpha_m = \frac{\beta_m [A]^m}{\Phi} ,$$

(I.25)

or, in logarithmic form:

$$\log \Phi = \log \beta_m + m \log [A] - \log \alpha_m.$$

(I.25a)

If we differentiate this equation and make use of equation (I.23), we obtain

$$\bar{n} = m - \frac{\partial \log \alpha_m}{\partial \log [A]} .$$

(I.26)

For maximum co-ordination of the complex MA_m we have the condition:

$$\frac{\partial \log \alpha_m}{\partial \log[A]} = 0;$$

(I.27)

so that at the point of maximum co-ordination of the complex MA_m:

$$\bar{n} = m; \qquad\qquad\qquad (I.26a)$$

and from equations (I.25a) and (I.24) it follows that

$$\log a_m = \int (m - \bar{n}) \, d\log [A] + \log \beta_n + B. \qquad (I.28)$$

If in the determination of the instability (or stability) constants we measure experimentally the equilibrium concentration of the central ion or of the ligand or of one of the complexes, then in the first case the degree of complex-formation is readily obtained, in the second case the formation function, and in the third case the fraction of the given complex. The corresponding instability constants are then calculated, using the established relationships between these constants and the above functions.

Very often, in the measurement of the instability constants, it is necessary to determine the composition of the complexes formed. For this purpose use is made of the various methods of physico-chemical analysis, together with data on the equilibria in the systems. The question of the determination of the composition of complex compounds formed in aqueous solution is dealt with in sufficient detail in the literature. Many of the most important methods, for example, are described in a monograph by Babko [9]. We shall therefore omit further discussion of this topic.

Fairly precise methods have now been developed for the calculation of instability (or stability) constants using the formation function and the degree of complex-formation. It is better, however, to discuss these methods using definite numerical illustrations taken from actual examples. We shall therefore calculate in later pages the formation function and degree of complex-formation, and use these to determine the instability (or stability) constants, for particular practical cases (for obvious reasons we shall limit ourselves to those methods most widely used in actual practice).

REFERENCES

1. J. BJERRUM, Metal Ammine Formation in Aqueous Solutions, Copenhagen (1941); quoted in Chem. Abs., 35, 6527 (1941).

2. I. LEDEN, Z. phys. Chem., A, 188, 160 (1941).
3. S. FRONAEUS, Acta Chem. Scand., 4, 72 (1950).
4. K. B. YATSIMIRSKII, Zh. neorg. khim., 1, 412 (1956).
5. K. B. YATSIMIRSKII, Zh. anal. khim., 10, 94 (1955).
6. J. C. SULLIVAN and J. C. HINDMAN, J. Am. Chem. Soc., 74, 6091 (1952).
7. H. IRVIN and H. S. ROSSOTTI, J. Chem. Soc., 3397 (1953).
8. J. Z. HEARON and J. B. GILBERT, J. Am. Chem. Soc., 77, 2594 (1955).
9. A. K. BABKO, Physico-chemical Analysis of Complex Compounds in Solution (Fiziko-khimicheskii analiz kompleksnykh soyedinenii v rastvorakh), Kiev (1955).

C H A P T E R II

EXPERIMENTAL METHODS FOR THE DETERMINATION
OF INSTABILITY CONSTANTS

A large number of methods of the greatest diversity are now being used for the determination of instability constants. It is impossible for us to go into description of the detailed procedure for each experiment. These problems are dealt with in a number of special textbooks and in a whole series of monographs (on potentiometry, polarography, spectrophotometry, etc.), so that we shall for the most part concern ourselves only with methods for calculating the instability constants from experimental data.

It is convenient to divide the methods for determining the instability constants of complex particles into two main groups.

Group I consists of methods which make possible the direct determination of the equilibrium concentration of one or several of the types of particle taking part in reactions (I.1), (I.3) or (I.4) (solubility methods, potentiometric methods, etc.).

Group II comprises methods based on the calculation of the changes in the physico-chemical properties of a system, taking place as a result of complex-formation (change in optical density, electrical conductivity, etc.). Using these methods it is impossible to calculate directly from experiment the equilibrium concentrations of the components during stepwise complex-formation.

METHODS IN GROUP I

The methods in Group I may be subdivided as follows:

A. Methods Based on the Study of Heterogeneous Equilibria
   1. The solubility method - the determination of the solubility of a sparingly soluble salt in the presence of complex-
forming substances, or of the solubility of electrically
neutral ligands in aqueous solutions in which complex-formation is taking place. The substance going into solution
forms particles which take part in the complex-formation.

   2. The distribution method - the study of the distribution
of the central ion, the ligand or the complex particle between
two immiscible solvents (normally water and an organic solvent
such as CCl_4).

   3. The ion exchange method - the study of the distribution
of the central ion or the ligand between a solution and an
ion exchange resin (cation exchanger or anion exchanger).

In the determination of instability constants using these
methods, a quantitative study is made of the heterogeneous
equilibrium in which the central ion, ligand or complex take
part, to find the equilibrium constant for the heterogeneous
process. The numerical value of this constant is then used
to calculate the equilibrium concentrations of the components
in a given phase.

B. Electrometric Methods
   When electrometric methods are used, a study is made of the
equilibria between the free metal and its ions in solution,
or between ions of the same element with different degrees
of oxidation, in order to determine the equilibrium concentrations of these particles in solution. This group consists of the following methods:

   1. The potentiometric method, which can be used (with
choice of a suitable electrode) to measure the equilibrium
concentration of the central ion or ligand.

   2. The polarographic method, which enables the instability
constant to be determined from a comparison of the polarographic curves recorded in the presence and in the absence
of substances causing complex-formation.

C. Other Methods in the First Group
   1. The kinetic method is based on the measurement of the
rate of any given reaction in which one of the components of
the dissociation equilibrium of the complex particle takes

part, first in the presence and then in the absence of the complex-forming substances.   Since the rate of a reaction depends on the concentration of the reactants, the method makes it possible to determine experimentally the equilibrium concentration of the central ion, ligand or complex.

2. The "freezing" method.   This is used for the study of the equilibrium constants of complex-formation reactions which take place slowly.   It is convenient to determine the concentration of the dissociation products of the complex (or complex ions) using labelled atoms.   The essential feature of the method is that one of the substances taking part in the dissociation equilibrium of the complex is rapidly and quantitatively removed from the sphere of the reaction.

3. The colorimetric indicator method.   The equilibrium concentration of the reactants may also be determined by measuring the optical density of a solution in which a coloured compound is present in equilibrium with one of the particles taking part in equations (I.1) or (I.3).   The equilibrium concentration of iron ions, for example, may be found from the optical density of solutions containing thiocyanate ions;   or the equilibrium concentration of hydrogen ions may be found by studying the behaviour of coloured indicators.

4. The biological method.   This is based on the study of the influence of the equilibrium concentration of any given ion on the function of a particular organ in a living organism (the heart of a frog, for example).

The action of the particular organ in systems in which complex-formation is taking place may be used to determine the equilibrium concentration of the ion under study.

5. The radioactive indicator method involves the measurement of the rate of isotopic exchange of a cation between the simple aquo-ion and a complex ion in solution, and extrapolation of the data obtained to "instantaneous" exchange.

We shall now examine the above methods in more detail.

A. Methods Based on the Study of
Heterogeneous Equilibria

1. The Solubility Method

The dissolution of a sparingly soluble salt in excess precipitant was first observed at a very early date and has often
been used (and still is) as a qualitative indication that
complex-formation is taking place in a given system. At the
same time it should be noted that if dissolution in excess
precipitant does not take place, this does not mean that complex-formation is impossible. Very often the solubility
product of a sparingly soluble salt is much smaller than the
instability constant of the complex formed. In this case
practically no dissolution takes place, although it is still
possible to prove the existence of complexes.

The solubility method was resorted to in the first years of
the twentieth century [1, 2, 3] for the quantitative calculation of the instability constants of the complexes formed,
and it has since become firmly established as a practical
method for the study of complex compounds.

When equilibria involving complex ions are being studied by
the solubility method, the substances saturating the solution
(the insoluble phase) may be solid salts containing the complex-
forming metal ions, or electrically neutral ligands (solid,
liquid or gaseous), or complex salts, or salts whose anions
form the ligands.

In the case of a sparingly soluble salt of the type MX_p
in a saturated solution, we have an equilibrium defined by the
definite value of the solubility product:

$$MX_p \rightleftarrows M^{pz+} + pX^{z-} \tag{II.1}$$

The ligands A* present in solution combine with the central
ion to form a series of complexes of the type MA_i (MA,
$MA_2,..., MA_{n-1}, MA_n$).

It is assumed that the anion, X, of the salt MX_p does not
form complexes with M, and that mixed complexes of the type
MXA are not formed.

---

* For simplicity the charge is omitted.

The solubility of the salt MX_p, i.e., the total concentration of metal in the solution, will be the sum of the concentrations of the separate complex particles and the free metal:

$$S = (c_M + c_{MA} + c_{MA_2} + \ldots), \qquad (II.2)$$

where S is the solubility of the salt MX_ϱ in mole/l.;

c_M is the concentration of metal ions not bound in a complex; and

c_{MA}, c_{MA_2} etc., are the concentrations of the complexes MA, MA_2, etc.

The equilibrium concentration of M ions is found from the solubility product of the salt MX_p:

$$[M] = \frac{K_s}{[X]^p}. \qquad (II.3)$$

From this it is possible to calculate the degree of complex-formation from the relationship:

$$\Phi = \frac{c_M^0}{[M]} = \frac{S[X]^p}{K_s}. \qquad (II.4)$$

When the ligand is not the anion of the sparingly soluble salt, the concentration of X anions is related directly to the solubility:

$$x = pS \qquad (II.5)$$

from which

$$\Phi = \frac{p^p S^{p+1}}{K_s}. \qquad (II.4a)$$

When the degree of complex-formation is known, the intermediate stability constants may be calculated using Leden's method [4]. In this case the value of β_1 is found from the equation:

$$\psi_1 = \frac{\Phi - 1}{[A]} = \beta_1 + \beta_2[A] + \beta_3[A]^2 + \ldots \qquad (II.6)$$

The equation (II.6) is obtained from (I.16) by simple conversion.

The function ψ_1 becomes β_1 when $[A]=0$. In practice
the constant β_1 is found by graphical extrapolation of the
function $\psi_1 = f([A])$; it is given by the point of intersection
of the curve with the y-axis. The function ψ_2 is then found
from the equation:

$$\psi_2 = \frac{^r\psi_1 - \beta_1}{[A]} \qquad\qquad (II.7)$$

or

$$\psi_2 = \frac{\Phi - 1 - \beta_1\,[A]}{[A]^2} . \qquad\qquad (II.7a)$$

The function ψ_2 is obtained from $(II.6)$ by a method similar to that used to obtained ψ_1 from $(I.16)$. This function may be represented as follows:

$$\psi_2 = \beta_2 + \beta_3\,[A] + \ldots \qquad\qquad (II.8)$$

The value of β_2 is found by extrapolating ψ_2 to zero
ligand concentration.

The values of the other stability constants are found by
similar methods from the functions ψ_3, ψ_4 and in general ψ_n.

It is also possible to carry out successive differentiation
of the values of the degree of complex-formation and use the
derivative values to find the stability constants.

When the value of the degree of complex-formation is known,
therefore, the calculation of the intermediate stability constants presents no difficulty.

In order to calculate the degree of complex-formation, it
is necessary in turn to know the values of the solubility,
the solubility product and the equilibrium concentration of
the ligand. The calculation of the solubility product from
solubility measurements alone is not always possible. It
is normally necessary, in addition to the solubility determination, to carry out appropriate potentiometric or conducto-
metric measurements, by means of which it is possible to
determine accurately the degree of dissociation of the sparingly soluble salt. In certain particular cases, when, for
example, the instability constants of the particles formed
in solution when the sparingly soluble salt dissolves are
known, the solubility product may be calculated from the
measured solubility alone.

In the study of the stability of complexes by the solubility method, the equilibrium concentration of the ligand is
scarcely ever determined, although a knowledge of this quantity is very valuable and simplifies the calculation considerably, particularly when the solubility is of the same order
of magnitude as the ligand concentration. The equilibrium
concentration of the ligand is usually determined by one of
the following methods.

If the solubility of the sparingly soluble salt is much
less than the corresponding ligand concentration, it may be
assumed with sufficient accuracy that the equilibrium concentration of the ligand is equal to its original concentration.
If, however, the solubility of the salt and the concentration of the ligand are commensurable, and particularly when
the difference between these quantities is small, the method
of successive approximations is used.   There are a number of
specific variations of this method, but all are essentially
the same. In determining the equilibrium ligand concentration
by this method, it is first assumed that only one complex
exists, and the equilibrium ligand concentration is calculated
with rough approximation, using an equation of the type:

$$\bar{c}_a = c_a^0 - a S, \qquad (\text{II.9})$$

where $\bar{c}_a$ is the equilibrium concentration of the ligand,

c_a^0 is the total (gross) concentration of the ligand,

S is the solubility of the salt, and

a is a coefficient related to the co-ordination
   numbers of the complexes formed and the stoichiometric coefficients of the sparingly soluble
   salt under study.

The coefficient a increases with increase in the ligand concentration, since the proportion of highly-co-ordinated complexes in the solution is increased, i.e., the so-called
average co-ordination number increases.   As a first approximation, however, it is assumed that the coefficient a is
constant and it is given a numerical value dependent on the
actual conditions;   the equilibrium concentrations are then
calculated.   It can be shown that if the substance in the
solid phase has a formula of the type MA, and if complexes
MA_2, MA_3 etc. are formed on dissolution, then $a > 2$;   if
the formula of the sparingly soluble salt is of the type M_2A,
and complexes MA, MA_2 etc. are formed on dissolution, then

again $a \geqslant 2$, .   If the anion of the salt does not take part
in the complex-formation, then $a \geqslant 1$, irrespective of the
formula type of the salt.

Let us assume, for example, that when a sparingly soluble
salt MA   dissolves in solutions of a ligand A with concentration c_a, complexes MA_2, MA_3, MA_4 etc. are formed.   The
concentration of complexes of type MA is usually so small
that it can be neglected.   The equilibrium concentration of
the ligand $\bar{c}_a$ in the sense of the determination is the difference

$$\bar{c}_a = c_a^0 - (c_a)_{\text{bound}} \tag{II.10}$$

where c_a^0 is the total (gross) concentration of ligand in
      the solution, and

   $(c_a)_{\text{bound}}$ is the concentration of ligand bound in a
      complex.

The term $(c_a)_{\text{bound}}$ may be expressed as

$$(c_a)_{\text{bound}} = 2c_{MA_2} + 3c_{MA_3} + 4c_{MA_4} + \ldots \tag{II.11}$$

Since the solubility S is given by:

$$S = c_{MA_2} + c_{MA_3} + c_{MA_4}, \tag{II.12}$$

then equation (II.11) becomes

$$(c_a)_{\text{bound}} = 2S + c_{MA_3} + 2c_{MA_4}, \tag{II.11a}$$

and equation (II.10) becomes

$$\bar{c}_a = c_a^0 - 2S - c_{MA_3} - 2c_{MA_4}. \tag{II.10a}$$

As a first approximation, c_{MA_3} and $2c_{MA_4}$ are negligible compared with $2S$, so that we obtain finally:

$$\bar{c}_a = c_a^0 - 2S. \tag{II.10b}$$

This expression may be simplified if we use the relationship

$$c_a^0 = c_a + S, \tag{II.13}$$

where c_a is the original concentration of ligand.

We then have

$$\bar{c}_a = c_a - S. \qquad (\text{II.10c})$$

The values of the equilibrium concentrations obtained are used to calculate approximate values of the instability constants of the complexes formed, from which new values are obtained for the equilibrium concentrations. The instability constants are then recalculated and used to calculate new equilibrium concentrations, and the process is repeated until concordant results are obtained.

As an example we may carry out the calculation for the equilibria set up when AgCNS dissolves in potassium thiocyanate solutions. Cave and Hume [5] have described the experimental determination of the solubility of AgCNS in water and in potassium thiocyanate solutions over a wide range of concentrations at a constant ionic strength equal to 2.2. The solubility product of AgCNS at this ionic strength is 6.75 x 10^{-12}.

The equilibrium thiocyanate concentration necessary for the calculation of the degree of complex-formation is found to a first approximation using equation (II.10c).

Table 1 gives the results of the experimental determination of the solubility of AgCNS in thiocyanate solutions, the values of the equilibrium thiocyanate concentrations calculated as a first approximation using (II.10c), and the degree of complex-formation calculated using equation (II.4), again as a first approximation, in view of the approximate nature of the value found for the equilibrium thiocyanate concentration and the function ψ_n.

The instability constants can now be calculated from the numerical values for the corresponding equilibrium thiocyanate concentrations and the degree of complex-formation. Before proceeding to the automatic application of the above formulae, however, we shall find it useful to examine the specific features of the system under study. In the present case, for example, it is clear that the concentration of the complex AgCNS, whose stability is defined by the stability constant β_1, is very small and is independent of the thiocyanate concentration. This can be seen from the following simple calculation. Cave and Hume [5] found the solubility of AgCNS in water to be 1.1 x 10^{-6} mole/l., and the value

TABLE 1

Solubility of AgCNS in thiocyanate solutions

Original thiocyanate concentration c_{CNS^-} (mole/l.)	Solubility of AgCNS (S) (mole/l.)	Equilibrium thiocyanate concentration $\bar{c}_{CNS^-}$ (mole/l.)	Degree of complex-formation	ψ_2	ψ_3	ψ_4
0.00548	$1.62 \cdot 10^{-6}$	0.00548	$1.31 \cdot 10^{3}$	$4.37 \cdot 10$	$1.59 \cdot 10^{9}$	$3.47 \cdot 10^{10}$
0.01033	$3.65 \cdot 10^{-6}$	0.01033	$5.61 \cdot 10^{3}$	$5.23 \cdot 10^{7}$	$1.68 \cdot 10^{8}$	$2.70 \cdot 10^{10}$
0.04133	$3.00 \cdot 10^{-5}$	0.04130	$1.83 \cdot 10^{4}$	$1.07 \cdot 10^{8}$	$1.74 \cdot 10^{9}$	$8.25 \cdot 10^{9}$
0.04440	$3.36 \cdot 10^{-5}$	0.04437	$2.20 \cdot 10^{4}$	$1.12 \cdot 10^{8}$	$1.74 \cdot 10^{9}$	$7.67 \cdot 10^{9}$
0.06662	$7.99 \cdot 10^{-5}$	0.06654	$7.89 \cdot 10^{4}$	$1.78 \cdot 10^{8}$	$2.15 \cdot 10^{9}$	$1.12 \cdot 10^{10}$
0.08885	$1.39 \cdot 10^{-4}$	0.08871	$1.81 \cdot 10^{6}$	$2.32 \cdot 10^{8}$	$2.22 \cdot 10^{9}$	$9.25 \cdot 10^{9}$
0.1111	$2.38 \cdot 10^{-4}$	0.1109	$4.89 \cdot 10^{4}$	$3.17 \cdot 10^{8}$	$2.54 \cdot 10^{9}$	$1.03 \cdot 10^{10}$
0.1334	$3.56 \cdot 10^{-4}$	0.1330	$7.00 \cdot 10^{4}$	$3.96 \cdot 10^{8}$	$2.71 \cdot 10^{9}$	$9.86 \cdot 10^{9}$
0.1779	$7.24 \cdot 10^{-4}$	0.1772	$1.91 \cdot 10^{7}$	$6.05 \cdot 10^{8}$	$3.21 \cdot 10^{9}$	$1.02 \cdot 10^{10}$
0.2224	$1.28 \cdot 10^{-3}$	0.2211	$4.20 \cdot 10^{7}$	$8.40 \cdot 10^{8}$	$3.64 \cdot 10^{9}$	$1.01 \cdot 10^{10}$
0.2744	$2.21 \cdot 10^{-3}$	0.2722	$8.90 \cdot 10^{7}$	$1.20 \cdot 10^{9}$	$4.26 \cdot 10^{9}$	$1.05 \cdot 10^{10}$
0.2774	$2.28 \cdot 10^{-3}$	0.2751	$9.31 \cdot 10^{7}$	$1.23 \cdot 10^{9}$	$4.33 \cdot 10^{9}$	$1.06 \cdot 10^{10}$
0.3343	$3.70 \cdot 10^{-3}$	0.3306	$1.81 \cdot 10^{8}$	$1.61 \cdot 10^{9}$	$4.75 \cdot 10^{9}$	$1.02 \cdot 10^{10}$
0.4443	$8.26 \cdot 10^{-3}$	0.4360	$5.34 \cdot 10^{8}$	$2.82 \cdot 10^{9}$	$6.37 \cdot 10^{9}$	$1.13 \cdot 10^{10}$
0.5572	0.0146	0.5426	$1.17 \cdot 10^{9}$	$4.00 \cdot 10^{9}$	$7.30 \cdot 10^{9}$	$1.09 \cdot 10^{10}$
0.5536	0.0146	0.5390	$1.17 \cdot 10^{9}$	$4.02 \cdot 10^{9}$	$7.40 \cdot 10^{9}$	$1.11 \cdot 10^{10}$
0.7783	0.0376	0.7407	$4.13 \cdot 10^{9}$	$7.54 \cdot 10^{9}$	$1.01 \cdot 10^{10}$	$1.17 \cdot 10^{10}$
1.114	0.0981	1.016	$1.47 \cdot 10^{10}$	$1.44 \cdot 10^{10}$	$1.42 \cdot 10^{10}$	$1.26 \cdot 10^{10}$
1.688	0.2684	1.420	$5.64 \cdot 10^{10}$	$2.80 \cdot 10^{10}$	$1.97 \cdot 10^{10}$	$1.29 \cdot 10^{10}$
2.252	0.5061	1.746	$1.31 \cdot 10^{11}$	$4.30 \cdot 10^{10}$	$2.46 \cdot 10^{10}$	$1.33 \cdot 10^{10}$

of K_s for this salt, calculated from potentiometric measurements, to be 1.13×10^{-12} (in terms of the activities at zero ionic strength).

The Ag^+ concentration is equal to $\sqrt{K_s}$, and amounts to 1.06×10^{-6}, so that the concentration of undissociated molecules, determined as the difference $1.1 \times 10^{-6} - 1.06 \times 10^{-6}$, is smaller by at least one order of magnitude than the solubility at the lowest thiocyanate concentration used in the experiments.

The separate terms of the right-hand side of equation (II.6) are proportional to the concentrations of the corresponding complex particles, so that the term $\beta_1[CNS]$ may be neglected. The fact that the AgCNS concentration is negligibly small, even when compared with the figure 10^{-6}, makes it impossible to estimate β_1 from the available data. The function ψ_2 is therefore calculated directly as

$$\psi_2 = \frac{\Phi}{[CNS^-]^2} = \beta_2 + \beta_3 [CNS^-] + \dots \qquad \text{(II.7b)}$$

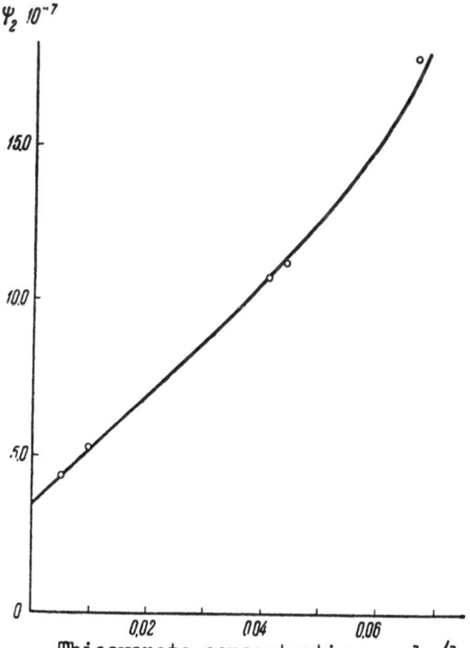

Fig. 1 Graphical determination of β_2

The data in Table 1 show that $\Phi \gg 1$ (as is generally the case in the vast majority of examples of the solubility method), so that $\Phi - 1 \simeq \Phi$, as has been assumed in (II.7b). The numerical values of the function ψ_2 are given in Table 1. Figure 1 gives the graph of ψ_2 vs. thiocyanate concentration; extrapolation to zero thiocyanate concentration gives the value $\beta_2 = 3.5 \times 10^{-7}$.

The next function ψ_3 is calculated from the formula:

$$\psi_3 = \frac{\psi_2 - \beta_2}{[CNS^-]} = \beta_3 + \beta_4 [CNS^-] + \ldots \qquad (II.7c)$$

Graphical extrapolation to $[CNS^-] = 0$ in Figure 2 gives the value $\beta_3 = 1.4 \times 10^9$.

Similar calculations for the function ψ_4 and extrapolation of the graph in Figure 3 gives the value $\beta_4 = 1.0 \times 10^{10}$.

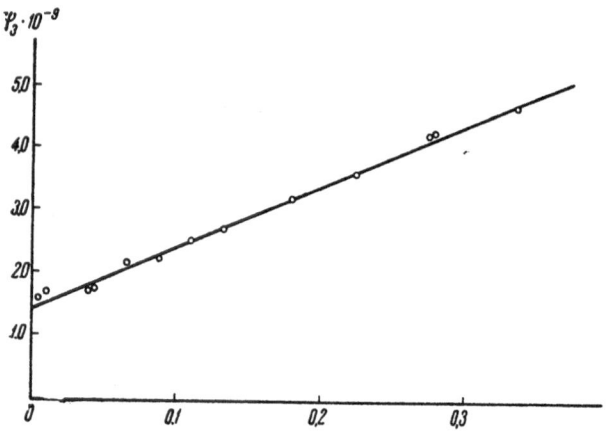

Thiocyanate concentration, mole/l.

Fig. 2.   Graphical determination of β_3

The approximate values of the stability constants can be used to carry out a more precise calculation of the equilibrium concentrations.   Continuing the derivation on page 20 we substitute in equation (II.10a) the concentrations of the complexes MA_3 and MA_4, expressed in terms of the

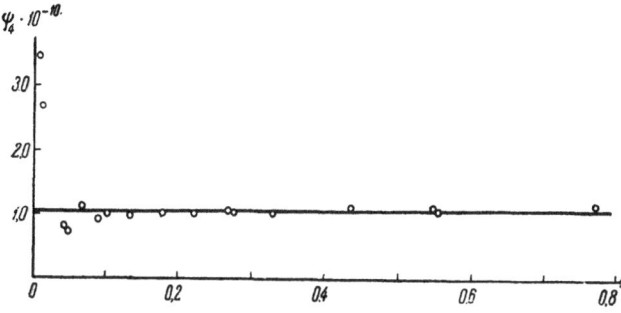

Thiocyanate concentration, mole/l.

Fig. 3. Graphical determination of β_4

corresponding stability constants and K_s for silver thiocyanate, i.e.:

$$c_{MA_3} = \beta_3 K_s \bar{c}_a^2; \; c_{MA_4} = \beta_4 K_s \bar{c}_a^3.$$

After substituting and making use of (II.13), we have:

$$\bar{c}_a = c_a - S - \beta_3 K_s \bar{c}_a^2 - 2\beta_4 K_s \bar{c}_a^3. \tag{II.10d}$$

Substituting the numerical values of β and K_s:

$$\bar{c}_a = c_a - S - 9.45 \cdot 10^{-3}\bar{c}_a^2 - 6.75 \cdot 10^{-2}\bar{c}_a^3. \tag{II.10e}$$

When the equation obtained is solved using Horner's method or the method of substitutions, values are obtained for the equilibrium concentrations. These values are used to carry out a more precise calculation of the stability constants.

In the study of the solubility of a sparingly soluble salt in the presence of excess anion, which functions in this case as a ligand, the solubility curve shows a minimum whose existence was experimentally established at an early date [6, 7]. The physical significance of the appearance of the minimum in this case is quite clear. The initial decrease in the solubility of the salt compared with the solubility in pure water is due to the common ion effect in accordance with the solubility product principle. The increase in the solubility with further increase in the ligand concentration is related only to complex-formation, if there are grounds for neglecting the salt effect, i.e., the decrease in the activity coefficients. If the minimum on the solubility curve ("the threshold of complex-formation") is fixed accurately, it is possible to calculate the instability constant from the following equation:

$$K_{\text{instab}} = (^n/_m q - 1) c_a^q, \tag{II.14}$$

where m and n are the stoichiometric coefficients of the sparingly soluble salt $A_m B_n$;

    q is the co-ordination number of the complex formed at the minimum point, and

    c_a is the concentration of ligand at the minimum point.

Equation (II.14) is obtained by differentiating a solubilit equation of type (II.12) and equating the derivative $\frac{\partial S}{\partial c_a}$ to zero. The detailed derivation of equation (II.14) and a number of examples of its application are given by Reynolds and Argensinger [8].

If a study is being made of the solubility of a salt MX_p in solutions containing M' ions, which form with X a series of complexes of the type $M'X_i$, it is more convenient to calculate the formation function. The total concentration of ligand in this case is calculated from the solubility of the salt, the equilibrium concentration of ligand from the solubility product of the salt MX_p, and the total concentration of metal is equal to the original concentration of $M'(c^0_{M'})$

$$\bar{n} = \frac{pS - \sqrt[p]{\dfrac{K_s}{[M]}}}{c^0_{M'}}. \tag{II.15}$$

The metal M should not form complex compounds with the anions X, and the concentration of M is thus practically equal to the concentration of the salt.

For this

$$\bar{n} = \frac{pS - \sqrt[p]{\dfrac{K_s}{S}}}{c^0_{M}}. \tag{II.15a}$$

When the solution is saturated with molecules of an electrically neutral ligand, its equilibrium concentration at constant temperature and constant ionic strength is also constant (this concentration may be denoted by S_0). In solutions containing ions of a metal M, some of the ligand molecules react to form complexes of the type MA_i and the solubility of the ligand increases to a value S.

From the solubility of the ligand in the pure solvent (S_0)
and its solubility in the presence of the complex-former (S)
it is an easy matter to calculate the formation function:

$$\bar{n} = \frac{S - S_0}{c_M},$$
                                                    (II.16)

where c_M is the total concentration of the ions of the
complex-forming metal M.

Andrews and Keefer [9], in a study of the solubility of
aromatic hydrocarbons in silver nitrate solutions, have
established the existence of complexes of the type $AgAr^+$
(Ar represents benzene, naphthalene, phenanthrene, toluene,
etc.) and have determined the instability constants of these
complexes.

More complex variations of the application of the solubility method are also possible. Vosburgh and Derr [10] have
studied the solubility of silver iodate in ammoniacal solutions containing nickel. From the solubility of silver
iodate and the instability constant of the complex ammines
of silver they have calculated the equilibrium concentration
of ammonia and hence the instability constants of the complex ammines of nickel.

2. The Distribution Method
The method involving study of the distribution of a substance between two immiscible solvents in the presence and in
the absence of complex-formation has since an early date been
applied to the determination of instability constants [2].

In the study of complex-formation by this method it is
possible to measure the distribution of the salt, the ligand,
or the complexes formed; but in practice the studies are
usually confined to the measurement of the distribution of
the metal salt between aqueous and nonaqueous phases. As a
rule, the distribution is studied at constant pH and in the

presence of a sufficient excess of a reagent HB which forms
an inner complex salt with the metal under study:

$$M_{aq}^{z+} + z\,HB_{solv} \rightleftarrows MB_{z\,solv} + z\,H_{aq}^{+}. \tag{II.17}$$

The subscripts aq and $solv$ denote particles present in the
aqueous solution and in the organic solvent respectively.
The equilibrium constant for this process is:

$$K = \frac{[MB_z]_{solv}\,[H^+]_{aq}^z}{[M^{z+}]_{aq}\,[HB]_{solv}^z}. \tag{II.18}$$

If the concentrations of H^+ and HB are constant, as
assumed above, we have

$$\frac{[MB_z]_{solv}}{[M^{z+}]_{aq}} = R_0, \tag{II.18a}$$

where R_0 is the distribution coefficient in the absence of
complex-formation.   The ligand A may be introduced into the
aqueous phase, whereupon the ratio of the concentrations of
metal in the organic solvent layer and in the aqueous layer
is altered:

$$\frac{[MB_z]_{solv}}{c_{M_{aq}}} = R, \tag{II.18b}$$

where $c_{M_{aq}}$ is the total concentration of the metal in
the aqueous phase, and

R is the distribution coefficient when complex-
formation takes place.

From equation (I.15), therefore, it follows that the degree
of complex-formation is given by the relationship

$$\Phi = \frac{R_0}{R}. \tag{II.19}$$

Day and Stoughton [11] have determined the instability constants of a whole series of complexes of thorium, by studying the extraction equilibrium for thorium in the presence
and in the absence of the various ligands.   It was assumed
that in the absence of complex-formation the thorium is
present in the aqueous phase in the form Th^{4+}.   The nonaqueous phase consisted of a solution in benzene of thenoyltrifluoroacetone, which reacts with thorium to form a complex

ThT_4, soluble in benzene.

The extraction equilibrium may therefore be represented:

$$Th^{4+} + 4HT_b \rightleftarrows (ThT_4)_b + 4H^+$$

and defined by the equilibrium constant:

$$K = \frac{[ThT_4]_b \cdot [H^+]^4}{[Th^{4+}] \cdot [HT]^4_b};$$

where the index b denotes the benzene phase.

On the introduction of a substance HX (for example, HF) whose anion $X^-(F^-)$ is able to form complexes with thorium, the following equilibria are set up in the aqueous phase:

$$Th^{4+} + HF \rightleftarrows ThF^{3+} + H^+$$

$$Th^{4+} + 2HF \rightleftarrows ThF_2^{2+} + 2H^+$$

etc., each of which is defined by an equilibrium constant

$$k_1 = \frac{[ThF^{3+}] \cdot [H^+]}{[Th^{4+}] \cdot [HF]},$$

$$k_2 = \frac{[ThF_2^{2+}] \cdot [H^+]^2}{[Th^{4+}] \cdot [HF]^2}.$$

It can be shown that the distribution coefficient of thorium in this case is given by the expression:

$$P = \frac{[Th^{4+}]}{[ThT_4]_b} \left(1 + k_1 \frac{[HF]}{[H^+]} + k_2 \frac{[HF]^2}{[H^+]^2} + \cdots \right),$$

and, in the absence of complex formation, by

$$P_0 = \frac{[Th^{4+}]}{[ThT_4]_b},$$

the expression in curved brackets being the degree of complex formation Φ, so that:

$$\Phi = \frac{P}{P_0} \; .$$

If we have as ligand the anion of a strong acid, for example NO_3^-, the calculations are simplified, since in this case the equilibrium concentration of the ligand is independent of the pH of the solution.

3. The Ion Exchange Method

If an ion exchange resin, for example a cation exchanger containing the cation $M_1^{z\,+}$, is placed in a solution containing $M_2^{r\,+}$ ions, exchange of ions will take place between the resin and solution until equilibrium is established:

$$r\,M_1R_z + z\,M_{2\,aq}^{r+} \rightleftharpoons z\,M_2R_r + r\,M_{1\,aq}^{z+}. \qquad (\text{II.20})$$

The equilibrium constant in this case is given by the equation:

$$K = \frac{[M_{1\,aq}^{z+}]^r\,[M_{2\,(R)}^{r+}]^z}{[M_{2\,aq}^{r+}]^z\,[M_{1\,(R)}^{z+}]^r} \; . \qquad (\text{II.21})$$

The subscript aq here denotes the ions present in the aqueous solution, while the subscript (R) denotes the ions in the cation exchanger.

If there is now introduced into the solution a ligand A which reacts with one of the ions to form a complex which is not adsorbed by the cation exchanger, the equilibrium (II.20) will be disturbed. The concentrations of ions in the solution (or in the cation exchanger) can be determined by analysis and used, together with the previously calculated equilibrium constant, to find the equilibrium concentrations of all the reacting particles.

The calculation is carried out in exactly the same way when an anion exchanger is used and a study made of the exchange between the anions in solution and the anions in the resin. The corresponding equilibrium constant is calculated, the concentrations determined by analysis after introduction of the metal, and the constant used to calculate the equilibrium concentrations.

It is possible to calculate either the degree of complexformation or the formation function, depending on which ion (metal ion or anion) is adsorbed by the ion exchange resin.

An example very frequently encountered in practice is the study of complex-formation by doubly- and triply-charged central ions with a singly-charged anion or neutral ligand. In this case the solution contains several particles carrying charges of the same sign, for example positively charged M^{z+}, MA^{z-1} , etc., which may be adsorbed by the cation exchanger.

A method for determining the composition and instability constants of the complex ions formed in solution in this case was proposed by Fronaeus [12-14].

If a study is made of the complexes formed by a doubly-charged complex-forming ion M^{2+} and a singly-charged ligand anion, then according to Fronaeus the ion exchange may be expressed by the equations:

$$2RNa + M^{2+} \rightleftarrows MR_2 + 2Na^+, \tag{II.22}$$

$$RNa + MA^+ \rightleftarrows MAR + Na^+; \tag{II.23}$$

each of which is defined by an equilibrium constant:

$$K_1 = \frac{[MR_2][Na^+]^2}{[RNa]^2[M^{2+}]}, \tag{II.22a}$$

$$K_2 = \frac{[MAR][Na^+]}{[RNa][MA^+]}. \tag{II.23a}$$

The expression for the distribution coefficient in this case is more complicated:

$$y = \frac{c_{MR}}{c_M} = \frac{[MR_2] + [MAR]}{c_M}. \tag{II.24}$$

The calculation of the instability constants is also more involved.

A description of the calculation may be found in the papers by Fronaeus cited above, or in a review article by Fomin [15]. The latter has also examined the possible applications and the most important limitations of the method.

B. Electrometric Methods

1. The Potentiometric Method

The potentiometric method is one of the oldest methods for determining instability constants. It is still being used with much success, and several variations of the method are known.

Instability constants may be measured by the potentiometric method using metal and amalgam electrodes, electrodes of the second type, redox systems, the glass electrode, etc.

The oldest modification of the potentiometric method consists of the measurement of the e.m.f. of the concentration cell:

$$M \mid M^{z+} R^+ X^- \quad \begin{vmatrix} M^{z+} \\ MA_n \end{vmatrix} M.$$

One of the compartments in this cell contains a solution of a salt of the metal under study, while the other contains the same solution with ligand added. The electromotive force (E) of the cell is equal to:

$$E = \frac{RT}{nF} \ln \frac{c_M}{[M]}, \qquad (II.25)$$

where c_M is the total concentration of the metal salt (the same in both compartments), and
[M] is the equilibrium concentration of metal ions in the compartment containing the complex-forming species.

The quantity whose logarithm appears in equation $(II.25)$ is the degree of complex-formation Φ, so that:

$$\ln \Phi = \frac{nF}{RT} E. \qquad (II.26)$$

The logarithm of the degree of complex-formation and the e.m.f. of the concentration cell are directly proportional to one another.

At present considerably wider practical use is made of a method involving potentiometric measurement of the pH and hence the equilibrium concentration of ligand (with the condition that the latter exhibits acid-base properties). The procedure in this case involves one of the various methods of potentiometric titration. For example, the solution to be titrated may consist of a salt of a complex-forming metal in acid medium, to which a solution of the ligand (a base) is added from a burette, while measurements are made of the pH. The acid-base dissociation constant of the ligand is previously determined under the same conditions of temperature and ionic strength; the total (analytical) concentrations of metal and ligand are known from the experimental conditions, and the pH of the solution is determined experimentally. The above data can be used to calculate the formation function and the equilibrium concentration of ligand. These calculations become somewhat more complicated as the basic strength of the ligand increases, but if the dissociation constants are known a fairly strict solution is possible and the results obtained are satisfactory.

The methods for finding the successive instability constants from the known formation function have been the subject of fairly numerous studies [17, 18, 25], and cannot be discussed here. We shall confine ourselves to a more or less detailed example of the calculation as applied in the determination of the formation function and instability constants of the propylenediamine complexes of nickel. The paper from which this example is taken is by Carlson, McReynolds and Verhoek [16].

The reaction between the nickel ion and propylenediamine (pn) leads to the formation of complexes of the type $Ni(pn)_n^{2+}$, where n may have the values 1, 2 and 3, depending on the propylenediamine concentration. The experiment was carried out as follows: 50.19 ml. of a solution containing 0.5 M potassium chloride, 0.1009 M hydrochloric acid and 0.04725 M nickel chloride was treated with an accurately measured volume of 4.822 M propylenediamine and the pH of the solution measured. The temperature was maintained at 30°.

The total propylenediamine concentration $|c_{pn}^0|$ is given by the sum:

$$c_{pn}^0 = [pn] + [pn\,H^+] + [pn\,H_2^{2+}] + \bar{n}c_{Ni^{2+}}, \qquad (II.27)$$

where $[pn]$, $[pnH^+]$ and $[pnH_2^{2+}]$ are the equilibrium concentrations of, respectively, the electrically neutral

propylenediamine, the product of the addition
of one proton from the hydrochloric acid, and
the product of the addition of two protons
from the hydrochloric acid;

$\overline{n}$ is Bjerrum's formation function, defined by
equation (II.32), and

$c_{Ni}{}^{2+}$ is the total concentration of nickel in the
solution.

We determine α, the ratio of the equilibrium concentration
of propylenediamine molecules to the concentration of pn,
which is not bound in a complex, i.e.,

$$\alpha = \frac{[pn]}{[pn] + [pnH^+] + [pnH_2^{2+}]} =$$

$$= \frac{k_{pnH^+} \cdot k_{pnH_2^{2+}}}{k_{pnH^+} \cdot k_{pnH_2^{2+}} + k_{pnH_2^{2+}} \cdot [H^+] + [H^+]^2},$$ (II.28)

where

$$k_{pnH^+} = \frac{[H^+][pn]}{[pnH^+]} \quad \text{and} \quad k_{pnH_2^{2+}} = \frac{[H^+][pnH^+]}{[pnH_2^{2+}]}.$$

We find the average number of hydrogen ions $(\overline{n}_A)$ attached
to a propylenediamine molecule not bound in a complex:

$$\overline{n}_A = \frac{[pnH^+] + 2[pnH_2^{2+}]}{[pn] + [pnH^+] + [pnH_2^{2+}]} =$$

$$= \frac{k_{pnH_2^{2+}}[H^+] + 2[H^+]^2}{k_{pnH^+} \cdot k_{pnH_2^{2+}} + k_{pnH_2^{2+}}[H^+] + [H^+]^2}.$$ (II.29)

Combining (II.27) and (II.29), we obtain for the formation
function:

$$\overline{n} = \frac{c_{pn}^0 - [pn] - [pnH^+] - [pnH_2^{2+}]}{c_{Ni}^{2+}} =$$

$$= \frac{c_{pn}^0 - \dfrac{c_{H^+} - [H^+]}{\overline{n}_A}}{c_{Ni}^{2+}}.$$ (II.30)

An expression for the calculation of the equilibrium propylenediamine concentration can be found by combining equations (II.28) and (II.29).

$$[pn] = \frac{\alpha\,([pn\,H^+] + 2\,[pnH_2^{2+}])}{\bar{n}_A} = \frac{\alpha}{\bar{n}_A}\,(c_{H^+} - [H^+]). \qquad (II.31)$$

If the initial concentrations of propylenediamine, hydrochloric acid and nickel salt are known, and the pH of the solution is determined experimentally, it is possible to calculate α and $\bar{n}_A$ using formulae (II.28) and (II.29) and then, using formulae (II.30) and (II.31) respectively, to calculate the formation function and the equilibrium propylenediamine concentration. The acid dissociation of the propylenediamine is measured beforehand under the same conditions of temperature and ionic strength.

The necessary experimental material and the results of the calculations are given in Table 2, borrowed from [16].

TABLE 2

Formation function $\bar{n}$ for propylenediamine-nickel complexes (temperature 30°; ionic strength 0.5)

Initial concentrations (mole/l.)			pH	$\alpha \cdot 10^5$ from equation (II.28)	$\bar{n}_A$ from equation (II.29)	$\bar{n}$ from equation (II.30)	$- \log$ [pn]
HCl	NiCl$_2$	pn					
0.0997	0.04669	0.05677	5.00	0.0132	2.000	0.148	8.18
0.0995	0.04659	0.06679	5.25	0.0414	1.984	0.356	7.68
0.0994	0.04652	0.07472	5.38	0.0733	1.978	0.527	7.43
0.0990	0.04633	0.0937	5.61	0.213	1.965	0.937	6.97
0.0987	0.04615	0.1136	5.89	0.746	1.938	1.337	6.42
0.0982	0.04597	0.1305	6.16	2.46	1.888	1.708	5.89
0.0978	0.04577	0 1511	6.56	13.7	1.763	2.091	5.12
0.0974	0.04561	0.1671	6.89	47.5	1.597	2.326	4.54
0.0971	0.04544	0.1848	7.14	115	1.458	2.601	4.11
0.0967	0.04526	0.2031	7.40	259	1.312	2.859	3.72
0.0965	0.04518	0.2108	7.56	427	1.233	2.933	3.48

The formation function $\bar{n}$ is related to the stability constants of the complexes by the well-known equation:

$$\bar{n} = \frac{x_1 [pn] + 2x_1 x_2 [pn]^2 + 3x_1 x_2 x_3 [pn]^3}{1 + x_1 [pn] + x_1 x_2 [pn]^2 + x_1 x_2 x_3 [pn]^3} . \qquad (II.32)$$

Solving this equation for the separate stability constants, we find:

$$x_1 = \frac{1}{[pn]} \cdot \frac{\bar{n}}{(1 - \bar{n}) + (2 - \bar{n}) [pn] x_2 + (3 - \bar{n}) [pn]^2 x_2 x_3} , \qquad (II.33)$$

$$x_2 = \frac{1}{[pn]} \cdot \frac{\bar{n} - 1) + \dfrac{\bar{n}}{[pn] x_1}}{(2 - \bar{n}) + (3 - \bar{n}) [pn] x_3} , \qquad (II.34)$$

$$x_3 = \frac{1}{[pn]} \cdot \frac{\bar{n} - 2) + \dfrac{\bar{n} - 1}{[pn] x_2} + \dfrac{\bar{n}}{[pn] x_1 x_2}}{(3 - \bar{n})} . \qquad (II.35)$$

These equations may be reduced to simpler, approximate formulae if it is assumed that at the given ligand concentration only two particles MA_{n-1} and MA_n are present in the solution. It can be seen that if in this case $[MA_{n-1}] = [MA_n]$, then $\bar{n} = n - 1/2$ and as a first approximation:

$$x_n = \left(\frac{1}{[pn]}\right)_{\bar{n} = n - 1/2} \qquad (II.36)$$

The ligand concentrations for the values $\bar{n} = 1/2, \ 1^1/2, \ 2^1/2$ may be used to calculate the stability constants $x_1, \ x_2, \ x_3$. Substituting (II.36) in (II.33), (II.34) and (II.35) we obtain:

$$x_1 = \frac{1}{[pn]_{\bar{n}=1/2}} \cdot \frac{1}{1 + 3x_2 [pn]_{\bar{n}=1/2} + 5x_2 x_3 [pn]^2_{\bar{n}=1/2}} , \qquad (II.33a)$$

$$x_2 = \frac{1}{[pn]_{\bar{n}=1/2}} \cdot \frac{1 + \dfrac{3}{x_1 [pn]_{\bar{n}=-1/2}}}{1 + 3x_3 [pn]_{\bar{n}=1/2}} , \qquad (II.34a)$$

$$x_3 = \frac{1}{[pn]_{\bar{n}=1/2}} \left(1 + \frac{3}{x_2 [pn]_{\bar{n}=1/2}} + \frac{5}{x_1 x_2 [pn]_{\bar{n}=1/2}}\right) . \qquad (II.35a)$$

If the stability constants of the complexes differ considerably from one another, the approximate formula (II.36) gives good results used alone.

The normal procedure for finding $\varkappa$ by this method is as follows.

A series of values of [pn] corresponding to $\bar{n} = 1/2$, $3/2$, and $5/2$ are chosen on the "formation curve" (Bjerrum's terminology), i.e., on the graph of $\bar{n}$·vs($-$ log [pn]) ; equations (II.33a), (II.34a) and (II.35a) then enable $\varkappa$ to be found.

The formation curve is shown in Figure 4, while Table 3 gives the results of the calculation of the stability constants.

TABLE 3

Stability constants of propylenediamine-nickel complexes

Calculated from the formation curve according to formula (II.36)			Calculated from equation (II.33a), (II.34a) and (11.35a)		
log $\varkappa_1$	log $\varkappa_2$	log $\varkappa_1$	log $\varkappa_1$	log $\varkappa_2$	log $\varkappa_1$
7.48	6.23	4.27	7.41	6.30	4.29

The data in Table 3 show that in this case the equation (II.36) by itself gives sufficiently accurate results.

This method is not the only one by which the stability constants may be found from the formation function.

Fronaeus' method [17], for example, involves graphical integration of the formation function in accordance with equation (I.24) and calculation of the degree of complex-formation. The stability constants are then calculated using Leden's method [4]. We have already discussed the application of Leden's method (page 17). A number of other methods are also known [18, 19].

Poulsen, Bjerrum and Poulsen [20] have recently put forward a simple method for finding the successive stability constants from the formation function. Their argument is as follows.

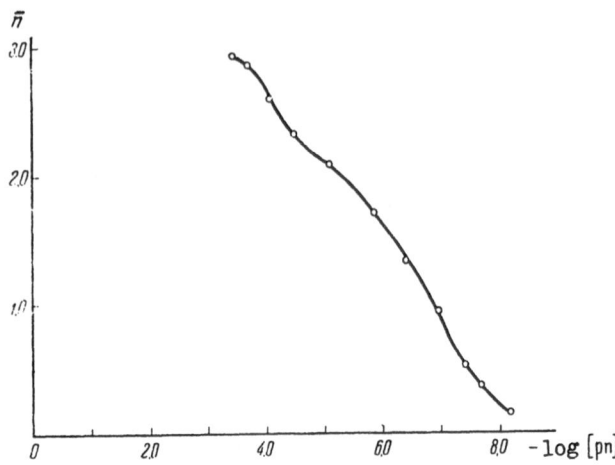

Fig.4 Formation curve for the Ni^{2+}
—propylenediamine (pn) system

We divide the formation function by the equilibrium ligand concentration:

$$\frac{\bar{n}}{[A]} = \frac{\beta_1 + 2\beta_2\,[A] + 3\beta_3\,[A]^2 + \ldots}{1 + \beta_1\,[A] + \beta_2\,[A]^2 + \beta_3\,[A]^3 + \ldots} \qquad (II.37)$$

and let

$$\frac{\bar{n}}{[A]} = f. \qquad (II.38)$$

It is obvious that f approaches β_1 as the ligand concentration approaches zero, i.e.,

$$\lim_{[A]\to 0} f \to \beta_1. \qquad (II.39)$$

Extrapolation of the function to $[A]=0$ in the system of co-ordinates $f - [A]$ gives β_1.

Transforming the expression (II.37) and taking (II.38) into account, we obtain:

$$\frac{f - \beta_1}{[A]} = \frac{2\beta_2 + 3\beta_3\,[A] - \beta_1^2 - \beta_1\cdot\beta_2\,[A] - \beta_1\cdot\beta_3\,[A]^2 + \ldots}{1 + \beta_1\,[A] + \beta_2\,[A]^2 + \beta_3\,[A]^3 + \ldots}. \qquad (II.40)$$

It is clear that:

$$\lim_{[\Lambda] \to 0} \frac{f - \beta_1}{[\Lambda]} = 2\beta_2 - \beta_1^2.$$
(II.41)

From which, similarly, we may find $2\beta_2 - \beta_1^2$ and then β_2.

By repeating the construction of functions analogous to (II.40) and carrying out graphical extrapolation, we find all the successive constants. The method is somewhat similar to the method of Leden described above.

When the various extrapolation methods for finding the constants (for example Leden's method) are used, it is necessary to check the accuracy of the constants obtained. The method of checking consists essentially in the substitution of the numerical values obtained for the constants in the fundamental equation relating the instability (or stability) constants, the equilibrium concentrations (of ligand or metal) and the property measured experimentally in the study of complex-formation (solubility, potential, etc.).

For example, if the extrapolation method is used in the treatment of data obtained in solubility studies, a comparison is made of the experimentally measured solubility and the solubility calculated using the constants obtained. When the constants found are accurate, the discrepancy between the calculated and experimental solubility values is usually found to be within the limits of the possible experimental error, the deviations being statistical in character.

Sufficient attention is not always paid to this feature, although it is absolutely necessary to check the constants in this way.

2. The Polarographic Method
The polarographic and potentiometric methods have much in common, since they are both based on the study of the interaction between an electrode and ions present in solution.

In the polarographic method, the equilibrium concentration of the metal (central ion) is determined from the shift in the half-wave potential when various concentrations of ligand are added to the test solution. The method has been used fairly widely, particularly in recent years. The application of the polarographic method is limited to the study of

complexes formed by cations which are reduced at a mercury
cathode. The complex compounds should be formed and break
down instantaneously (i.e., should be "reversible complexes",
to use Heyrovsky's terminology).

Heyrovsky [21] was the first to use polarographic curves
for the calculation of instability constants. The equation
for the displacement of the half-wave potential on complex-
formation was subsequently made more precise, and is now
applied in the following form (for 25°):

$$(E_{1/2})_{complex} - (E_{1/2})_0 =$$

$$= \frac{0.059}{n} \log K - \frac{0.059}{n} \log \frac{f_M k_k}{f_k k_M} - \frac{0.059}{n} p \log c_x f_x, \qquad (II.42)$$

where $(E_{1/2})_{complex}$ and $(E_{1/2})_0$ are the values of the half-
   wave potentials in solutions of the complex salt
   (with ligand added) and the simple salt respectively.

K is the instability constant,

k_k and k_M are the proportionality coefficients
   relating concentration and current strength
   for the complex and simple ions respectively,

f_M, f_k and f_x are the activity coefficients of
   the metal ions, complex ions and ligands x.
   respectively, and

p is the co-ordination number of the complex formed.

Equation (II.42) is usually simplified by making a number
of approximately justifiable assumptions, such as:

$$k_K = k_M, \text{ and } f_M = f_k,$$

which give:

$$(E_{1/2})_{complex} - (E_{1/2})_0 = \frac{0.059}{n} \log K - \frac{0.059}{n} p \log c_x f_x. \quad (II.42a$$

Sometimes f_x is taken as equal to unity, which simplifies
equation (II.42a) still further.

Many workers have used equation (II.42a) in this form for

the calculation of instability constants. By means of this
equation we can determine the co-ordination number (from the
gradient of the tangent to the curve $\Delta E_{1/2}$ vs.$\log c_x$) and instability constant of the complex formed.   Equation (II.42)
has been derived with the assumption that only one complex,
with co-ordination number p, is formed in solution.   In
actual fact, however, the solution contains several complexes
with different co-ordination numbers, so that the curve
$\Delta E_{1/2}$ vs.$\log c_x$ is not a straight line but is convex towards the
x-axis.   It is very difficult, and sometimes impossible, to
calculate from the equation the equilibrium constants for the
intermediate stages of complex-formation.   If the intermediate instability constants of the separate complex particles
differ sufficiently from one another, it is sometimes poss-
ible to distinguish straight regions on the curve $\Delta E_{1/2}$ vs.$\log c$,
which then has the form of a broken line.   It is then poss-
ible to determine p and $\log K$ by applying equation (II.42a)
to each section of this broken line.

Yatsimirskii [22] has used a logarithmic graph method to
calculate the successive constants.   Assuming that two complexes exist at each ligand concentration, he obtained the
equation:

$$\Delta E_{1/2} = \frac{0.059}{n} \log K - \frac{0.059}{n} (p-1) \log c_x - \frac{0.059}{n} \log \left(c_x + \frac{1}{k}\right) \cdot \text{(II.43)}$$

Differentiation gives

$$\frac{\partial \Delta E_{1/2}}{\partial \log c_x} = -\frac{0.059}{n} \left(p - 1 + \frac{c_x}{c_x + k}\right). \qquad \text{(II.44)}$$

At the points where $c_x = k$, the tangent satisfies the condition that:

$$\frac{\partial \Delta E_{1/2}}{\partial \log c_x} = -\frac{0.059}{n} (p - 1/2). \qquad \text{(II.44a)}$$

By drawing a series of tangents satisfying condition (II.44a)
at various values of p, we find the co-ordination numbers
and instability constants of the complexes formed in solution.

The successive instability constants can also be calculated
using Leden's method.   The application of this method to
polarographic measurement of instability constants has been

described by DeFord and Hume [23]. The authors showed that
the shift in the half-wave potential is related to the degree
of complex-formation:

$$\log \Phi = 0.435 \frac{nF}{RT} \Delta E_{1/2} + \log \frac{J_1}{J_2},$$ (II.45)

where J_1 is the diffusion constant of the simple ion;

   J_2 is the apparent (average) diffusion constant of
   the series of complex ions. The ratio of these
   constants is usually equal to 1, so that the
   second term on the right-hand side of equation
   (II.45) disappears;

$\Delta E_{1/2} = (E_{1/2})_0 -- (E_{1/2})_{complex}$ is the shift in the half-
   wave potential on complex-formation.

Substituting the constant quantities in (II.45) we finally
obtain (for 25°):

$$\log \Phi = 17.0 n \Delta E_{1/2}.$$ (II.45a)

Hume, DeFord and Cave [24] have used this method to determine the instability constants of thiocyanato complexes of
cadmium. All their experiments were carried out at 30°.
Under these conditions formula (II.45) becomes:

$$\log \Phi = 34.5 \Delta E_{1/2}.$$ (II.45b)

The determination of the successive stability constants when
the degree of complex-formation is known has already been
discussed.

C. Other Methods in the First Group

1. The Kinetic Method
When one of the substances taking part in the dissociation
equilibrium for a complex particle

$$MA_n \rightleftarrows M + nA,$$

can react at a measurable rate with a third substance, this
reaction can be used to measure the instability constant.

If the ligand A reacts at a measurable rate with a substance B, then the rate of the corresponding reaction, for
example, the bimolecular reaction:

$$A + B \rightarrow D$$

is given by the equation

$$-\frac{dc}{dt} = kc_A c_B.$$

By measuring the rate of reaction at a known concentration of substance B, having previously determined the rate constant k, we can calculate the equilibrium concentration of the substance A, c_A. Thus, for example, the equilibrium concentration of thiosulphate ion may be determined from the rate of reduction of trivalent iron ions [26], and the equilibrium concentration of trivalent iron ions from the rate of oxidation of iodide [27].

Cases of particular value are those where the substance to be determined functions as a catalyst in a homogeneous catalysis reaction. An example is the study by Bell and Prue [28] of the rate of depolymerization of diacetone alcohol $CH_3COCH_2C(CH_3)_2OH \rightarrow 2CH_3COCH_3$. This is a catalytic reaction, the rate of which is proportional to the hydroxyl ion concentration. By studying this reaction in the presence of various cations, the authors determined the instability constants of the hydroxo complexes $CaOH^+$ and $BaOH^+$.

When kinetic methods are used to determine instability constants, the experimental data can be used for the direct calculation of either the degree of complex-formation or the formation function. If the equilibrium concentration of the metal is determined experimentally, it is more convenient to calculate the degree of complex-formation; if the equilibrium concentration of the ligand is known, it is more convenient to calculate the formation function.

The applications of the kinetic method are restricted by the fact that no sufficient study of the kinetics of appropriate reactions has been made, although such reactions could be suggested for a large number of ions. A discussion of the reactions which have been most studied, with a classification of the applications of the kinetic method, is to be found in a recent review [29].

2. The "freezing" method (determination of equilibrium concentrations by chemical analysis)
   The "freezing" method is widely used for the study of vapour-phase equilibria. The method can also be applied to the study of the instability constants of complexes whose rate of formation and breakdown is low. The substance to

be determined is removed rapidly and quantitatively from the
sphere of reaction by precipitation or by binding in an undissociated compound.   The concentration of the test substance determined in this way is the equilibrium concentration, since the systems chosen for study by this method are
those in which the equilibrium is established and disturbed
at a low rate.

This method has been used, for example, to determine the
instability constants of thiocyanato complexes of trivalent
chromium [20].

Most complexes with a slow rate of formation and breakdown
have an exceptionally high stability.   The concentration of
their breakdown products is therefore extremely low and can
only be measured using special methods.   It is particularly
convenient in this case to use labelled atoms.

Cook and Long [30] have determined the instability constants
of the iron complex ferroin and were the first to use labelled
atoms for this purpose.   The instability constants were determined by studying the equilibrium

$$Fe(Ph)_3^{2+} + 3H^+ \rightleftarrows Fe^2 + 3HPh^+.$$

A solution containing radioactive iron was prepared and
left for two days; non-radioactive $FeSO_4$ solution was then
added, after which the complex cation was precipitated with
a solution containing CdI_4^{2-} ions.   The Fe^{2+} ions which were
not combined in the complex were left in solution.   Measurements were made of the radioactivity of the original solution
and of the solution left after precipitation of the ferroin
complex.

The instability constant is readily calculated from the
data obtained and the known original concentrations of iron
salt and ferroin.

In the further development and application of this method
it is necessary to take into account the results obtained by
Grinberg and Nikol'skii [31], who have shown that even in
the case of very stable complexes exchange of radioactive
isotopes between the ions in solution and the ions in the
complex particle takes place fairly rapidly.   At the same
time reliable results can be obtained only when exchange is
very slow.

3. The Indicator Colorimetric Method

The equilibrium concentration of the reacting substances can also be determined from the optical density of a solution containing a coloured compound in equilibrium with one of the products of dissociation of the complex. Thus, for example, the equilibrium concentration of trivalent iron ions may be found from the optical density of solutions containing thiocyanate ions [32]. The equilibrium concentration of hydrogen ions may be found from a study of the behaviour of coloured indicators in the test solution. In the first case it is possible to calculate the degree of complex-formation, and in the second case the experimentally determined values of the pH of the solution may be used to calculate the equilibrium ligand concentration (provided the ligand exhibits acid-base properties) and from this the formation function.

The indicator colorimetric method makes possible the quantitative study of colourless complexes in the visible region of the spectrum. The basis of this type of determination is the "decolorization" of a coloured complex when new ligands or complex-forming species are added to the solution.

Babko and Rychkova [33] have determined the instability constant of the salicylato-aluminium complex from the discharge of the colour of the salicylato-iron complex when an aluminium salt is added to the solution:

$$FeSal^+ + Al^{3+} \rightleftarrows AlSal^+ + Fe^{3+};$$

$$K_p = \frac{[AlSal^+][Fe^{3+}]}{[FeSal^+][Al^{3+}]} = \frac{K_{FeSal^+}}{K_{AlSal^+}}.$$

The equilibrium constant for this process is determined experimentally, K_{FeSal^+} is known beforehand, and from these data it is possible to calculate K_{AlSal^+}.

When a solution of a ligand, such as fluoride, is added to this solution, the corresponding complexes of iron, in this case the fluoro complexes, are formed; a decrease in the optical density will again take place.

Data obtained in this way are also used for the calculation of instability constants.

4. The Biological Method
   In their determination of the instability constant of the
citrato complex of calcium, Hastings and co-workers [34] made
use of the effect produced by calcium ions on the ventricle
contraction of the isolated heart of a frog. The biological
action is shown only by calcium ions which are not bound in
a complex, which makes it possible to determine the equilibrium concentration of these ions in a solution. The un-
certainty associated with the activity of the heart was elim-
inated by comparing the effects produced by the test solution
and by a standard solution; the concentration of the standard solution was so selected that the amplitude of contraction was the same in both cases.

   This method is not of general interest in view of its
limited application, but for the study of certain systems it
has proved indispensable. Its use is most effective in
cases where the usual methods for determining free ion concentrations are inapplicable, for example in various types
of biological systems containing considerable quantities of
natural buffering agents.

5. The Radioactive Indicator Method
   Radioactive indicators have been widely used in recent
years in many varied fields of chemistry, but the method has
not been used for the determination of instability constants
to any great extent.

   On page 44 mention is made of the work of Cook and Long
[30] who used labelled atoms in the "freezing" method.

   Recently, Cook and Long [35] have determined instability
constants using a modification of the labelled atom method,
involving the measurement of the rate of isotopic exchange of
a cation between aquo-ions and complex ions in solution.

   By studying the exchange over varying periods of time and
extrapolating to zero time intervals ("instantaneous exchange"),
it is possible to find the equilibrium concentration of free
metal ions. This method has been used to study the complexes
formed by iron with ethylenediaminetetraacetate [36]. The
method can only be applied when the rate of exchange between
the aquo-ions and complex ions is sufficiently low.

METHODS IN GROUP II

Solutions of complex compounds exhibit properties which are essentially different from the sum of the properties of the original reagent solutions. The methods in the second group, in some form or other, make use of this deviation from additivity, which is related to the quantity, composition and properties of the new complex particles formed in the solution.

The second group of methods includes the following:

1. The spectrophotometric method, based on the study of the optical density and absorption spectra of the solutions.

2. The electrical conductivity method. When new complex ions are formed, the electrical conductivity of the system shows a departure from additivity. The extent of the deviation from additivity can be used to follow the process of complex-formation and to calculate the instability constants.

3. The cryoscopic and ebullioscopic methods, based on the study of the change brought about in the freezing point or boiling point of solutions by the change in the number of particles in solution during complex-formation.

4. The calorimetric method allows the number of complex particles formed in solution to be estimated from the magnitude of the thermal effect produced when the original solutions are mixed.

Let us examine the methods of Group II in more detail.

1. The Spectrophotometric Method

If one or more coloured complexes are formed in a system, the optical density of the solution will change with change in the ligand concentration. It is impossible to determine the equilibrium concentration of the complexes directly from the change in the optical density of the solution, since it is necessary to know the molar extinction coefficients of every complex formed in the system.

If the complex is sufficiently stable and its formation practically complete at high ligand concentrations, it is possible to find its molar extinction coefficient from these conditions or, in the last resort, to determine it by extrapolation of several values obtained at high ligand concentrations.

More complicated examples of the determination of the extinction coefficient are discussed in Babko's monograph [37].

A method for the calculation of extinction coefficients has been developed by Komar' [38, 39] for those cases where the colour of the solution is produced not only by the complex particles formed, but also by the reacting particles themselves. He has also examined cases where complex-formation in solution is accompanied by side reactions, such as hydrolysis, acid-base interaction, etc.

If one coloured compound only is formed in the solution, the calculation of the equilibrium constant is comparatively simple.

Considerable difficulty arises when several complexes are formed in the system. The calculation of instability constants in stepwise complex-formation from study of the physico-chemical properties of the solutions (including optical density) has been discussed by Yatsimirskii [40].

From a series of determinations of the optical density of solutions containing complex compounds, it is possible to obtain a series of values of the mean molar extinction coefficients according to the formula:

$$\bar{\varepsilon} = \frac{D}{C_M l}, \tag{II.46}$$

where $\bar{\varepsilon}$ is the mean molar extinction coefficient,

D is the optical density of the solution,

l is the thickness of the light-absorbing layer and

c_M is the total concentration of metal ions.

The ligands A present in the solution react with the ions of the metal to form a number of complex particles $MA_1, MA_2, \ldots MA_n$, with stabilities defined by the respective stability constants $\beta_1, \beta_2, \ldots, \beta^n$.

According to the Bouguer-Lambert-Beer law, the optical density of the solution may be expressed by the equation:

$$\frac{D}{l} = \varepsilon_0 [M] + \varepsilon_1 [MA] + \varepsilon_2 [MA_2] + \ldots + \varepsilon_n [MA_n], \tag{II.47}$$

where ε_0, ε_2, ..., ε_n are the molar extinction coefficients of M, MA, MA$_2$,..., MA$_n$ respectively.

From the obvious relationship:

$$c_M = [M] + [MA] + [MA_2] + \ldots + [MA_n] \qquad (II.48)$$

and equations (II.46) and (II.47), we have:

$$\bar{\varepsilon} = \frac{\varepsilon_0 + \varepsilon_1\beta_1 [A] + \varepsilon_2\beta_2 [A]^2 + \ldots + \varepsilon_n\beta_n [A]^n}{1 + \beta_1 [A] + \beta_2 [A]^2 + \ldots + \beta_n [A]^n} . \qquad (II.49)$$

If the value of ε_0 is subtracted from both sides of equation (II.49), we obtain:

$$\Delta\bar{\varepsilon} = \frac{\Delta\varepsilon_1\beta_1 [A] + \Delta\varepsilon_2\beta_2 [A]^2 + \ldots + \Delta\varepsilon_n\beta_n [A]^n}{1 + \beta_1 [A] + \beta_2 [A]^2 + \ldots + \beta_n [A]^n} , \qquad (II.50)$$

where $\Delta\bar{\varepsilon} = \bar{\varepsilon} - \varepsilon_0$; $\Delta\varepsilon_1 = \varepsilon_1 - \varepsilon_0$; $\Delta\varepsilon_2 = \varepsilon_2 - \varepsilon_0$; $\Delta\varepsilon_n = \varepsilon_n - \varepsilon_0$.

Equation (II.50) is true not only for the molar extinction coefficients, but also for the optical densities, if all the measurements are made in a cell with the same length of light-absorbing layer.

By carrying out a series of determinations, it is possible to obtain a large number of values for $\Delta\bar{\varepsilon}$ and hence a large number of equations of type (II.50). The problem is to discover methods for finding the coefficients in these equations $(\Delta\varepsilon, \beta)$.

The method proposed by Yatsimirskii for the solution of these equations involves the construction of a series of subsidiary functions and extrapolation of these to zero value of the variable.

Let us consider the simple subsidiary function

$$f_1 = \frac{\Delta\bar{\varepsilon}}{[A]} . \qquad (II.51)$$

From equations (49) and (50), it follows that:

$$f_1 = \frac{\Delta\varepsilon_1\beta_1 + \Delta\varepsilon_2\beta_2 [A] + \Delta\varepsilon_3\beta_3 [A]^2 + \ldots + \Delta\varepsilon_n\beta_n [A]^{n-1}}{1 + \beta_1 [A] + \beta_2 [A]^2 + \beta_3 [A]^3 + \ldots + \beta_n [A]^n} . \qquad (II.52)$$

Extrapolation of f_1 to zero ligand concentration gives:

$$\lim_{[A]\to 0} f_1 = a_1 = \Delta\varepsilon_1\beta_1. \tag{II.53}$$

This extrapolation may be carried out graphically by plotting the equilibrium ligand concentrations as abscissae against the values of f_1. (calculated from equation (II.51)) as ordinates. The point of intersection of the Y-axis then gives the value of $\Delta\varepsilon_1\beta_1$.

Differentiation of the function f_1 and extrapolation of the derivative to zero ligand concentration gives:

$$\lim_{A\to 0} \frac{df_1}{d[A]} = a_2 = \Delta\varepsilon_2\beta_2 - \Delta\varepsilon_1\beta_1^2. \tag{II.54}$$

The value of a_2 may also be found by constructing a new subsidiary function f_2:

$$f_2 = \frac{f_1 - a_1}{[A]}. \tag{II.55}$$

On extrapolation to zero ligand concentration, this function becomes:

$$\lim_{[A]\to 0} f_2 = a_2 = \Delta\varepsilon_2\beta_2 - \Delta\varepsilon_1\beta_1^2. \tag{II.56}$$

Similarly a third subsidiary function f_3 may be constructed:

$$f_3 = \frac{f_2 - a_2}{[A]}. \tag{II.57}$$

On extrapolation of f_3 to zero ligand concentration we obtain:

$$\lim_{[A]\to 0} f_3 = a_3 = \Delta\varepsilon_3\beta_3 - \Delta\varepsilon_1\beta_1^3. \tag{II.58}$$

The same result may be obtained by differentiating f_2 and extrapolating the derivative to zero ligand concentration.

Subsidiary functions f_4, f_5, f_6, etc., may be constructed in exactly similar fashion; in general –

$$f_i = \frac{f_{i-1} - a_{i-1}}{[A]}. \tag{II.59}$$

Extrapolation of this function to zero ligand concentration gives

$$\lim_{[A]\to 0} f_i = a_i = \Delta\varepsilon_i\beta_i - \Delta\varepsilon_1\beta_1^i. \tag{II.60}$$

Since we have to determine not only the values of the stability constants β_1, β_2, ..., β_n, but also the values of the changes in the molar extinction coefficients ($\Delta\varepsilon_1$, $\Delta\varepsilon_2$, ..., $\Delta\varepsilon_n$), the total number of extrapolation equations is equal to half the number of unknown quantities to be determined. In this connexion it is convenient to introduce a new variable y, related to the ligand concentration by the simple equation:

$$y = \frac{1}{[A]}. \tag{II.61}$$

Division of the numerator and denominator in the right-hand side of equation (II.50) by $[A]^{n'}$ and substitution from (II.61) gives:

$$\Delta\bar{\varepsilon} = \frac{\Delta\varepsilon_n\beta_n + \Delta\varepsilon_{n-1}\beta_{n-1}y + \ldots + \Delta\varepsilon_1\beta_1 y^{n-1}}{\beta_n + \beta_{n-1}y + \ldots + \beta_1 y^{n-1} + y^n}. \tag{II.62}$$

Extrapolation of $\Delta\bar{\varepsilon}$ to zero value of y gives:

$$\lim_{y\to 0} \Delta\bar{\varepsilon} = b_1 = \Delta\varepsilon_n. \tag{II.63}$$

Differentiation of $\Delta\bar{\varepsilon}$ with respect to y and extrapolation of the derivative to zero value of y gives the following result:

$$\lim_{y\to 0} \frac{d\Delta\bar{\varepsilon}}{dy} = (\Delta\varepsilon_{n-1} - \Delta\varepsilon_n)\frac{\beta_{n-1}}{\beta_n}. \tag{II.64}$$

An analogous result may also be obtained by constructing the subsidiary function φ_1:

$$\varphi_1 = \frac{\Delta\bar{\varepsilon} - b_1}{y}. \tag{II.65}$$

On extrapolation of this function to zero value of y, as can be seen from equation (II.62) we find that:

$$\lim_{y\to 0} \varphi_1 = b_2 = (\Delta\varepsilon_{n-1} - \Delta\varepsilon_n)\frac{\beta_{n-1}}{\beta_n}. \tag{II.66}$$

The functions φ_2, φ_3, ..., φ_{n-1} are then constructed in similar fashion and extrapolated to zero value of y.

By combining the values obtained by extrapolation of the
functions $\varphi_1, \varphi_2, \ldots, \varphi_{n-1}$, with the values obtained by extrapolation of the functions $f_1, f_2, f_3, \ldots, f_n$, we can find all
the coefficients of equation (II.50).

For example, when only two complexes MA and MA_2 are present
in solution, we have

$$a_1 = \Delta\varepsilon_1\beta_1,$$
$$a_2 = \Delta\varepsilon_2\beta_2 - \Delta\varepsilon_1\beta_1^2,$$
$$b_1 = \Delta\varepsilon_2,$$
$$b_2 = (\Delta\varepsilon_1 - \Delta\varepsilon_2)\frac{\beta_1}{\beta_2}.$$

This system of equations may be solved very simply, to
give, in addition to the stability constants β_1 and β_2, their
optical characteristics $\Delta\varepsilon_1$ and $\Delta\varepsilon_2$.

The values for the coefficients in equation (II.50) found
by this method require further confirmation, and this is
done by comparison of the calculated values of $\Delta\bar{\varepsilon}$ with the
experimental values over the whole range of ligand concentrations studied.

In the case of very stable complexes, the calculation of
the equilibrium ligand concentration presents certain diffi-
culties. If, however, the complexes are not extremely
stable, the total ligand concentration will not differ apprec-
iably from the equilibrium concentration.

The above method has been used by Yatsimirskii and Fedorova
[41] to calculate the stability constants of acetate complexes of divalent chromium. A number of variations of
equilibrium calculations from spectrophotometric data are
given in Babko's monograph [37].

2. The Electrical Conductivity Method
One of the first electrical conductivity methods was used
by Chernyayev and Khorunzhenkov [42] in the study of platinum
complexes.

This method was subsequently used by many research workers
to study the instability constants of a variety of acido-complexes.

In the electrical conductivity method, the conductivity of solutions of the pure salts M_nA and RX_m is first studied, after which the solutions are mixed and the conductivity of the resultant solution is measured again. If a complex RA^{m-n} is formed in the solution according to the scheme:

$$R^{m+} + A^{n-} \rightleftarrows RA^{m-n},$$

then the conductivity of the mixture will be less than that which would be observed in the solution in the absence of complex-formation, by a certain quantity equal to

$$(\varkappa_0 - \varkappa) 10^3 = an\lambda_A + am\lambda_R - a\lambda_{RA}. \qquad (II.67)$$

In this equation $\varkappa_0$ is the conductivity of the solution in the absence of complex-formation (calculated from conductivity data for the pure salts M_nA and RX_m), $\varkappa$ is the experimentally observed conductivity λ_A, λ_R, and λ_{RA} are the equivalent conductances of the ions A^{n-}, R^{m+} and RA^{m-n} respectively, n and m are the charges on the ions A^{n-} and R^{m+}, and a is the molar concentration of the complex RA^{m-n}. The ionic conductances for varying electrolyte concentration may be estimated from the equation:

$$\lambda_A = \lambda^0_A - B^{!:}$$

where λ^0_A is the ionic conductance at infinite dilution, and

B is a coefficient which may be determined by experimental study of the conductivities of the pure salts.

The ionic conductance of the complex RA^{m-n} is zero where m and n are equal, so that the complex carries no charge. It is usually assumed that the ionic conductance of this particle is simply related to the conductance of the anion A^{n-} (in the case of anionic complexes) or to that of the cation R^{m+} (in the case of positively charged complexes):

$$\lambda_{RA} = \frac{n - m}{n} \lambda_A \qquad (II.69)$$

or

$$\lambda_{RA} = \frac{m - n}{m} \lambda_R. \qquad (II.70)$$

Substitution of the appropriate values in equation (II.67) enables a to be found, after which the instability constant

Summary of methods of determination of instability constants

Method	No. of constants determined by the method *	Conditions of Applicability	Remarks
Solubility method	89	Presence of an insoluble phase of sufficiently low solubility, which increases sharply on complex-formation	The solid phase may consist of the simple salt, the ligand or the complex salt
Distribution method	32	Presence of a solvent capable of reversibly extracting the salt of the given metal from aqueous solution	The method is complicated by simultaneous extraction of ligand or complex compound
Ion exchange method	33	Reversible exchange of ions between the solution and the exchanger	
Potentiometric method	968	a) The existence of an electrode reversible with respect to the ions of the metal or ligand b) When the pH method is used – the presence of distinct acid-base reaction by the ligand	
Polarographic method	63	Reversible and instantaneous discharge of the complex ions at a mercury cathode	

Method	No.	Description	
Kinetic method	23	Presence of a reaction taking place at a measurable rate and involving the complex-forming ions or ligands (as reactant or catalyst)	
"Freezing" method	2	Slow formation and dissociation of the complexes	
Indicator colorimetric method	Included under spectrophotometric method	The presence of a reversible reaction involving the formation of a coloured compound, one of the reactants or products of which may form complexes with the complex-forming ions or ligand	
Biological method	1	The existence of a quantitative relationship between the functioning of a living organ and the concentration of an ion forming the complex	
Spectrophotometric method	98	A change in optical density on complex-formation	
Electrical conductivity method	60	A change in electrical conductivity on complex-formation	
Other methods	12		Not applicable to the formation of complexes with neutral ligands

* This column includes only those constants given in our Tables. The other constants given in the supplementary literature are not included here.

may be calculated from the equation:

$$K = \frac{(c_R^0 - a)(c_A^0 - a)}{a},$$ (II.71)

where c_R^0 and c_A^0 are the total (initial) concentrations of the ions R^{m+} and A^n

The electrical conductivity method has not yet been applied to systems in which several complexes are formed.

3. Cryoscopic and Calorimetric Methods
These methods have not been widely applied to the study of complex compounds. The first of them involves the calculation of the change in the number of particles in the system from the freezing point depression [43].

The equilibria in solutions in which complex-formation is taking place may also be studied by measuring the heats of mixing for different ligand concentrations. The essential features of the method generally employed are similar to those of the spectrophotometric method (page 47). In certain more simple cases, for example when only two types of particle are present in the solution, the calculations are simplified considerably [44].

The present chapter includes (pp. 54-55) a summary indicating (statistically) the extent to which the various methods for determining instability constants are employed, together with the chief limitations to their applicability.

REFERENCES

1. G. BODLANDER, R. FITTIG, Z. phys. Chem., 39, 597 (1902).
2. H. MORZE, ibid., 41, 709 (1902).
3. H. EULER, Ber., 36, 2878 (1903).
4. I. LEDEN, Z. phys. Chem., A, 188, 160 (1941).
5. G. S. CAVE and D. N. HUME, J. Am. Chem. Soc., 75, 2893 (1953).
6. W. D. HARKINS, ibid., 33, 1807 (1911).
7. M. BARRE, Ann. Chem. Phys., 24, 145 (1911).
8. C. A. REYNOLDS and W. I. ARGENSINGER, J. Phys. Chem., 56, 417 (1952).
9. L. J. ANDREWS and R. M. KEEFER, J. Am. Chem. Soc., 71, 3644 (1949).
10. P. E. DERR and W. C. VOSBURGH, ibid., 65, 2408 (1943).
11. R. A. DAY, Jnr., and R. W. STOUGHTON, ibid., 72, 5662 (1950).

12. S. FRONAEUS, Acta Chem. Scand., 5, 859 (1951).
13. S. FRONAEUS, ibid., 6, 1200 (1952).
14. S. FRONAEUS, ibid., 7, 21 (1953).
15. V. V. FOMIN, Usp. khimii, 24, 1010 (1955).
16. G. A. CARLSON, J. P. McREYNOLDS and F. H. VERHOEK, J. Am. Chem. Soc., 67, 1334 (1945).
17. S. FRONAEUS, Acta Chem. Scand., 4, 72 (1950).
18. H. IRVING and H. S. ROSSOTTI, J. Chem. Soc., 3397 (1953).
19. J. T. EDSALL et al., J. Am. Chem. Soc., 76, 3054 (1954).
20. K. G. POULSEN, J. BJERRUM and J. POULSEN, Acta Chem. Scand., 8, 921 (1954).
21. J. HEYROVSKY, Polyarograficheskii metod (Polarography) (Translation into Russian), ONTI (1937).
22. K. B. YATSIMIRSKII, Collected Works on General Chemistry (Sbornik statei po obshchei khimii), I, p. 193 (1953).
23. D. D. DeFORD and D. N. HUME, J. Am. Chem. Soc., 73, 5321 (1951).
24. D. N. HUME, D. D. DeFORD and G. C. B. CAVE, ibid., 73, 5323 (1951).
25. J. BJERRUM, Metal Ammine Formation in Aqueous Solutions, Copenhagen (1941).
26. K. B. YATSIMIRSKII, Zh. anal. khim., 10, 339 (1955).
27. K. W. SYKES, H. J. FUDGE, J. Chem. Soc., 124 (1952).
28. R. P. BELL and J. E. PRUE, ibid., 362 (1949).
29. K. B. YATSIMIRSKII, Zavod. lab., 21, 1410 (1955).
30. C. M. COOK and F. A. LONG, J. Am. Chem. Soc., 73, 4119 (1951).
31. A. A. GRINBERG and L. E. NIKOL'SKAYA, Zh. prikl. khim., 22, 542 (1949).
32. A. K. BABKO and K. E. KLEINER, Zh. obshch. khim., 17, 1259 (1947).
33. A. K. BABKO and T. N. RYCHKOVA, ibid., 18, 1617 (1948).
34. A. B. HASTINGS et al., J. Biol. Chem., 107, 351 (1934).
35. C. M. COOK and F. A. LONG, Cited in J. Phys. Chem., 56, 25 (1952).
36. S. S. JONES and F. A. LONG, J. Phys. Chem., 56, 25 (1952).
37. A. K. BABKO, Physico-chemical Analysis of Complex Compounds in Solution (Fiziko-khimicheskii analiz kompleksnykh soyedinenii v rastvorakh), Kiev (1955).
38. N. P. KOMAR', Uch. zap Khar'kovsk. Gosuniv., 37; Trud. nauchno-issled. inst. khimii, 8, Khar'kov (1951).
39. N. P. KOMAR', Uch. zap. Khar'kovsk. Gosuniv., 54; Trud. khimich. fak. i nauchno-issled. inst. khimii, 12, Khar'kov (1954).
40. K. B. YATSIMIRSKII, Zh. neorg. khim., 1, 2306 (1956).
41. K. B. YATSIMIRSKII and T. I. FEDOROVA, Zh. neorg. khim., 1, 2301 (1956).

42. I.I. CHERNYAYEV and S. I. KHORUNZHENKOV, *Izv. inst. platiny*, No. 7, 98 (1929).
43. SUSHIL K. SIDDHANTA, *Chem. Abstr.*, **45**, 7317 (1949).
44. K. B. YATSIMIRSKII and V. P. VASIL'EV, *Zh. fiz. khim.*, **30**, 901 (1956).

C H A P T E R III

COMPLETE THERMODYNAMIC DESCRIPTION OF
COMPLEX-FORMATION REACTIONS IN SOLUTION

In order to reach accurate conclusions regarding the nature
of the forces operating within complex particles during their
formation in solution, we need to know the energy changes
accompanying the reactions in question. Instability constants are related directly only to the change in free energy.
At the same time it is necessary to know the enthalpy change
on complex-formation (the thermal effect) in order to calculate the entropy change on complex-formation, the change in
stability of the complex particles with change in temperature,
and the bond energy.

A knowledge of the entropy change during the reactions
permits estimation of certain factors determining the stability of complex compounds. It is a requisite of a thorough
study of the processes of complex-formation in solution that
there should be a complete thermodynamic characterization of
the reactions in question, i.e., at the very least, determination of the changes in the enthalpy (ΔH), entropy (ΔS) and
free energy (ΔG).

The choice of the standard state is of very great importance in the determination of changes in the principal thermodynamic functions. Most authors choose as standard state
a hypothetical molal solution with the properties of an in-
finitely dilute solution. This choice of standard state
has certain advantages (elimination of the effect of ionic
strength, unambiguity), but also certain disadvantages, associated with the necessity for extrapolation (not always accurate) of the data to zero values of the ionic strength. In

recent years, therefore, use has been made of a less hypothetical molar solution of given ionic strength $(\mu = \text{const})$*
With this choice of standard state the need to extrapolate
the values obtained for the various thermodynamic functions
disappears. Care must be taken to ensure that all determinations, calculations and comparisons of thermodynamic
functions are made with reference to the same standard state.

Thermodynamic functions may be used for description of the
overall reaction of complex-formation:

$$M + nA = MA_n, \tag{III.1}$$

and also for description of the separate stages of this process:

$$MA_{i-1} + A = MA_i. \tag{III.2}$$

Since reaction (III.1) may be represented as the sum of the
reactions of type (III.2), the thermodynamic functions describing reaction (III.1) may be represented as the sum of the
corresponding functions describing the reactions of type
(III.2). In this sense we may speak of total (or overall)
and stepwise (or partial) changes in enthalpy, entropy and
free energy.

The equilibrium data can be used directly to calculate the
total and partial free energy changes:

$$\Delta G \text{ total} = -RT \ln \beta, \tag{III.3}$$

$$\Delta G \text{ partial} = -RT \ln \varkappa. \tag{III.4}$$

Alternatively, taking account of the relationship between the
instability constant and the stability constant, we may write:

$$\Delta G \text{ total} = RT \ln K, \tag{III.5}$$

$$\Delta G \text{ partial} = RT \ln k. \tag{III.6}$$

In all such calculations it is necessary to bear in mind
the standard state chosen and to calculate all ΔG values

---

* In all cases the temperature is taken as 25°.

for the appropriate ionic strength.

The enthalpy changes for reactions (III.1) and (III.2) are given by the thermal effects for these reactions. They may be found by direct calorimetric measurement or indirectly from the isobar equation for the chemical reaction, using equilibrium constant data obtained at different temperatures.

The isobar equation for the reaction:

$$\frac{d \ln \beta}{dT} = \frac{\Delta H}{RT^2},$$ (III.7)

where T is the absolute temperature and R the gas constant, may be more conveniently written in the form:

$$\frac{d \log \beta}{d (1/T)} = \frac{\Delta H}{4.57}.$$ (III.8)

ΔH may be calculated by first constructing the graph with co-ordinates $1/T$, log β. If the temperature range studied

TABLE 4

Thermal effect in the formation of BaEdta^{2-} in solution

Temperature °C	$1/T$	pK	Temperature interval, °C	$T_1 \cdot T_2$	ΔH from equation (III.10), kcal/mole
0	$3\,66 \cdot 10^{-3}$	8.01	—	—	—
5	$3.59 \cdot 10^{-3}$	7.95	0—5	$7.60 \cdot 10^4$	— 4.2
10	$3\,53 \cdot 10^{-3}$	7.89	5—10	$7.88 \cdot 10^4$	— 4.3
15	$3.47 \cdot 10^{-3}$	7.84	10—15	$8.16 \cdot 10^4$	— 3.8
20	$3.41 \cdot 10^{-3}$	7.78	15—20	$8.48 \cdot 10^4$	— 4.7
25	$3.36 \cdot 10^{-3}$	7.73	20—25	$8.74 \cdot 10^4$	— 4.0
30	$3.30 \cdot 10^{-3}$	7.68	25—30	$9.04 \cdot 10^4$	— 4.2

is relatively small, the value of ΔH remains approximately constant and the graph of log β versus $1/T$ is a straight line. The gradient of this line is numerically equal to $-\frac{\Delta H}{4.57}$. If the graph obtained is not a straight line, the gradient of the tangent to the curve at the point corresponding to a temperature of $25°$ (or any other desired temperature) is measured.

It is also possible to carry out an approximate integration of equation (III.8) on the assumption that the value of ΔH remains constant over the temperature range from T_1 to T_2 ; we then have:

$$\log \frac{\beta'}{\beta'} = \frac{\Delta H \cdot (T_2 - T_1)}{4.57 \cdot T_1 \cdot T_2}, \qquad (III.9)$$

where β'' and β' are the values of the stability constants relating to temperatures T_2 and T_1 respectively.

The above equation may also be written in the slightly different form:

$$pK'' - pK' = \frac{\Delta H (T_2 - T_1)}{4.575 \cdot T_1 \cdot T_2}. \qquad (III.10)$$

Equations (III.9) or (III.10) may be used to calculate the thermal effect of the reaction (ΔH) without graphical construction.

If we choose a ten-degree range of temperature $(T_2 - T_1 = 10°)$ from 20° to 30° (mean temperature 25°), substitute these values in equation (III.10) and recalculate ΔH in kcal/mole, we obtain:

$$\Delta H = 40\ 6\ \ \Delta pK_{(20-30°)}.$$
$$(III.10a)$$

A thermal effect of 1 kcal/mole thus corresponds to a change in pK of 0.025 for a temperature change of 10°.

As an example, let us calculate the thermal effect of the reaction:

$$Ba^{2+} + Edta^{4-} = BaEdta^{2-}$$

(Edta $^{4-}$ represents the ethylenediaminetetraacetate anion).

The data given in Table 4 show that the values calculated for the thermal effect using equation (III.10) are not very accurate.

The foregoing method for the determination of thermal effects is, as mentioned above, indirect. Its use requires extremely accurate determination of the equilibrium constants of reactions (III.1) and (III.2), since these normally change

very little over a small temperature range. The determinations should be carried out for a number of precisely fixed
temperatures (at least three). The ΔH values obtained by
this method are generally less accurate than the values obtained by direct calorimetric measurements.

To determine the heats of reactions of type (III.1) or
(III.2), we have to carry out a series of calorimetric measurements of the heats of mixing solutions of a salt containing M
ions with a series of solutions containing varying concentrations of the ligand A. The heats of dilution of the original solutions of the salt of metal M and of the ligand A
must also be measured and the appropriate corrections made
to the values obtained for the heats of mixing. The corrected
values for the heats of mixing represent the mean values of
the heats of formation of the complexes and are related to
the values of the heats of formation by the following expression:

$$\Delta \bar{H} = a_1 \Delta H_1 + a_2 \Delta H_2 + a_3 \Delta H_3 + \ldots, \qquad (III.11)$$

where $\Delta \bar{H}$ is the corrected heat of mixing for one mole
    of metal;

$a_1, a_2, a_3 \ldots$ are the fractions of complexes of type
    MA_1, MA_2, $MA_3 \ldots$, respectively , and

$\Delta H_1, \Delta H_2, \Delta H_3 \ldots$ are the heats of the reactions
    of type (III.2) for the values of n equal
    to 1, 2, 3... respectively.

Taking account of the relationships derived above, we find
that:

$$\Delta \bar{H} = \frac{\beta_1 \Delta H_1 [A] + \beta_2 \Delta H_2 [A]^2 + \beta_3 \Delta H_3 [A]^3 + \ldots}{1 + \beta_1 [A] + \beta_2 [A]^2 + \beta_3 [A]^3 + \ldots}. \qquad (III.12)$$

Values of $\Delta H_1, \Delta H_2, \Delta H_3 \ldots$ are calculated from a series of
measurements of the heats of mixing (the number of such
measurements should be not less in any case than the number
of heats of reaction to be determined); all stability constants and equilibrium ligand concentrations should be known.
Checking of the ΔH_n values obtained, by substituting them
in equation (III.12), is necessary in all cases. This substitution is significant only if the number of heats of mixing found is greater than the number of heats of reaction of
type (III.2) to be determined.

As mentioned above, equation (III.12) may also be used to find stability constants from thermodynamic data.

The heats of complex-formation reactions found by the two different methods should coincide. Nevertheless, marked discrepancies may be observed in practice between the two sets of figures. For example, Table 5 gives values for the heats of ethylenediaminetetraacetato complexes, found by the above two methods. It can be seen from the Table that in some cases the ΔH values obtained differ not only in magnitude but also in sign.

The values of the free energy changes are related by a very simple expression to the stability constant values, so that all the features mentioned above governing the changes in the stepwise constants may be applied directly to the changes in ΔG. The ΔG values, as a rule, increase on stepwise complex-formation. Exceptions are found only in the cases of the complexes of mercury, cadmium and indium referred to above, and apparently in certain other cases with tetrahedral co-ordination of the ligand anions.

The changes in the stepwise heats of attachment of ligands are in most cases quite different in character. In the formation of ammine complexes, the stepwise heats of complex-formation remain approximately constant. In most cases of acido-complex formation, the enthalpy changes accompanying

TABLE 5

Comparison of the thermal effects determined
calorimetrically with those calculated
from the isobar equation

Equation for the reaction	Thermal effect, kcal/mole	
	determined calorimetrically	found from the isobar equation
$Mg^{2+} + Edta^{4-} = MgEdta^{2-}$	3.1	—2.9
$Ca^{2+} + Edta^{4-} = CaEdta^{2-}$	—5.8	—2.5
$Sr^{2+} + Edta^{4-} = SrEdta^{2-}$	—4.2	—4.1
$Ba^{2+} + Edta^{4-} = BaEdta^{2-}$	—5.1	—4.1

the attachment of each new ligand become increasingly negative; the attachment reaction becomes more exothermic in spite of the decrease in the stability of the complex particles formed.

The differing nature of the changes in ΔG and ΔH in step-wise complex-formation is explained by the part played by the entropy factor.

The entropy change in the above reactions is calculated using the familiar equation:

$$\Delta S = \frac{\Delta H - \Delta G}{T}. \tag{III.13}$$

When equation (III.13) is used, it must be ensured with special care that the same standard state is chosen for the determination of ΔH and ΔG, since the value of the ionic strength of the solution has a particularly marked effect on the entropy change.

The entropy changes on complex-formation may be classified as:

1) the changes (always negative) related to decrease in the number of particles on complex-formation, and

2) the changes (always positive) related to the dehydration of the reacting particles in the process of complex-formation. A ligand which occupies several co-ordination positions displaces several molecules of water, so that the attachment of a ligand of this type is associated with a larger positive entropy change than that for a ligand which occupies only one co-ordination position and which displaces only one molecule of water. Ions with a high electric charge are more extensively hydrated than ions with smaller charge, and this also leads to a more positive entropy change in complex-formation involving highly-charged ions.

All comparisons of changes in thermodynamic functions are most conveniently made for series of reactions of the same type, i.e., reactions in which the same number of the same ligands are attached to the central atom to form particles with the same spatial configuration.

The entropy change in these cases obeys the fairly simple relationship:

$$\Delta S = 0.1 L_M + B, \tag{III.14}$$

where L_M is the heat of hydration of the ion M, and

B is a constant for a given series of reactions
    of the same type.

In stepwise complex-formation the attachment of each new
particle is accompanied by a decreasing entropy change. This
leads to a decrease in the stability of the complexes as more
ligands are added.

C H A P T E R   IV

FACTORS DETERMINING THE STABILITY OF

COMPLEX COMPOUNDS IN SOLUTION

The stability of a complex particle (ion or molecule) in
solution is determined by the nature of the central atom and
the ligands. The most important characteristics of the
central atom, determining the stability of the complex compound, are the degree of oxidation (charge on the central
ion in the case of ionic complexes), the dimensions, and the
electronic structure. In the case of complexes with mon-
atomic ligands, stability is dependent on the same characteristics in the ligand (charge, radius and electronic structure). The strength of binding for ligand molecules and
polyatomic ions depends, in addition, on the nature of the
atoms directly linked to the central atom, and on the particular features of the structure of the ligand molecule (or
ion).

The influence of the charge on the central ion on the stability of complexes may be seen from a comparison of the change
in stability in various series of complex compounds with the
same number of the same ligands, but with varying charge on
the central ions.

The central ions chosen for comparison have approximately
equal radii and similar electronic structure. Examples of
such cation series are: 1. Na^+, Ca^{2+}, Y^{3+}, Th^{4+} ($r_c = 1.04 \pm 0.06$ Å);
2. K^+, Sr^{2+}, La^{3+} ($r_c = 1.27 \pm 0.06$ Å); 3. Ag^+, Hg^{2+},
Tl^{3+} ($r_c = 1.09 \pm 0.04$ Å).

Table 6 gives the instability constant exponents for these three ion series. In all cases the value of pK increases with increase in the charge on the central ion. The change in the pK value with change in the charge is particularly sharp in those cases where the ligands are small or highly-charged ions (oxalate, ethylenediaminetetraacetate, etc.). In the case of large ions of small charge (iodide, bromide, nitrate, iodate, etc.), the change in stability is much less marked.

TABLE 6

Instability constant exponents in various series of ions with varying charge

Type of complex ion	pK			
	Charge on the central ion			
	+1	+2	+3	+4
K^+, Sr^{2+}, La^{3+}				
ROH^{z-1}	(—0.7)	1.0	3.3	—
$RS_2O_3^{z-2}$	0.1	2.0	—	—
$RP_3O_9^{z-3}$	1.2	3.4	5.7	—
$RFe(CN)_6^{z-3}$	1.2	2.8	3.7	—
Na^+, Ca^{2+}, Y^{3+}, Th^{4+}				
RJO_3^{z-1}	—	0.9	—	2.9
RNO_3^{z-1}	—	0.3	—	0.6
RSO_4^{z-2}	0.7	2.3	3.5	4.1
$RC_2O_4^{z-2}$	—	3.0	7.3	—
$REdta^{z-4*}$	—	11.1	18.0	—
Ag^+, Hg^{2+}, Tl^{3+}				
ROH^{z-1}	2.3	10.3	14.8	—
RNH_3^{z}	3.2	8.8	—	—
RCl^{z-1}	2.7	5.3	8.1	—
RBr^{z-1}	.9	9.1	9.7	—

* $Edta^{1-}$ denotes the ethylenediaminetetraacetate ion.

The change in pK with change in charge in the case of central ions with a complete electronic octet is so regular that in some cases it is possible to find unknown instability constants by interpolation.

It might be thought that the stability of complexes should increase smoothly with increase in charge, and that the most stable complexes should be formed by the most highly charged ions. In fact, highly charged ions form compounds with oxygen anions extremely readily as a result of interaction with water molecules (VO_2^+, MoO_2^{2+}, UO_2^{2+}, BiO^+ etc.). Since O^{2-} anions have a comparatively high charge and small size, the stability of such particles increases very rapidly with increase in the charge on the central ion, and exceeds that of all other possible complexes.

The stability of some complex compounds in the series Cu^+, Zn^{2+}, Ga^{3+} and Ag^+, Cd^{2+}, In^{3+} shows a decrease, in spite of the increase in charge. This type of change in stability can hardly be explained entirely by the change in radius and polarizing action of the central ions. It will be shown below that these and similar exceptions from the above rule are observed in cases where π-bonds are formed inside the complex particle between the ligands and the central atom.

We have shown from theoretical considerations [1] that with change in the radius of the central ion, the stability of complex compounds may increase, decrease or pass through a maximum. This may be seen particularly clearly by examining complexes of magnesium, calcium, strontium and barium.

TABLE 7

Change in stability in various series of
complex compounds formed by the elements
of Group 2 of Mendeleyev's periodic system

Formula of complex	pK values for complexes of type			
	MgX^{2-z}	CaX^{2-z}	SrX^{2-z}	BaX^{2-z}
MOH^+	2.58	1.30	0.82	0.64
MCH_3COO^+	0.82	0.77	0.44	0.31
$MCH_2ClNH_2COO^+$	1.96	1.24	—	0.77
MC_2O_4	3.43	3.0	2.54	2.33
$MP_3O_9^-$	3.31	3.47	3.35	3.25
$MP_4O_{12}^{2-}$	5.17	5.47	5.11	5.0
$MEdta^{2-}$	8.69	10.59	8.63	7.76
MNO_3^+	0.0	0.28	—	0.92
MJO_3^+	0.72	0.89	1.00	1.05
MS_2O_3	1.84	1.98	2.04	2.33

The examples given in Table 7 show that there is a certain optimum ratio of reacting particle dimensions in complex-formation: small cations form their most stable complexes with small anions, and large cations with large anions.   In the case of small singly-charged anions (anion radius less than 1.6 A.), α-aminoacids, and oxalate, the stability decreases with increase in cation size, i.e., from top to bottom of the appropriate group in the Mendeleyev Periodic System.

In the case of large anions (nitrate, iodate, thiosulphate, etc.), the stability of the complexes increases with increase in cation dimensions, and finally, in the case of certain triply-charged and quadruply-charged anions (tetrametaphosphate, ethylenediaminetetraacetate, tartrate, etc.), passes through a maximum, which normally appears at the calcium complex.   These rules are found to obtain so strictly that in a number of cases unknown instability constants can be calculated by interpolation and extrapolation [2].

The regular features examined here indicate that a direct link exists between the stability of complex compounds and the position of the elements in question in the Mendeleyev periodic system when these are in the main sub-groups, i.e., when they form ions with an inert gas type of electronic structure.   A similar relationship exists for the other elements, but quantitative consideration of this question is complicated by the fact that, within the complex compounds in question, the chemical bond is not purely ionic, and in a number of cases approximates to a purely covalent bond.   The stability of the covalent bond depends on a number of factors, and it is sometimes difficult to take complete account of these.

For complexes with the same number of the same ligands, formed by ions with the same charge and approximately the same volume, it is possible to derive the following semi-empirical relationship [2]:

$$pK = a + nbp, \tag{IV.1}$$

where a is a constant for a given series of ions with
   the same volume and charge,

   n is the number of ligands,

   b is the polarizability of the ligand, and

   p is a quantity defining the polarizing effect
   of the cation.

The term $nb\rho$ in this equation not only defines the polarization, in the restricted sense of the word (the displacement of the centres of gravity of the positive and negative charges), but also reflects the increase in the degree of covalency of the bond resulting from the displacement of the electrons from one of the ligand atoms to the central ion. In the limiting case, this displacement leads to the formation of a dative σ-bond.

As a first approximation, the polarizing action of ions varies in the same way as the ionization potentials of the electrically neutral atoms, which are numerically equal to the electron affinities of the ions in question, since the change in energy on the attachment of an electron to a gaseous doubly-charged ion is numerically equal to the second ionization potential:

$$M^{2+}_{(gas)} \dotplus \bar{e} = M^{+}_{(gas)}$$

The relationship between the polarizing action and the ionization potential should, however, be regarded as only approximate.

The simultaneous solution of equations of type (IV.1) for two series of complex compounds formed by ions of the same volume and charge makes it possible to eliminate the unknown polarizing action of the cation: the instability constant exponent for some well-studied series of complexes is taken as a standard. We have taken the polarizability of the alaninate ion as unity ($b = 1$), so that for complexes of the type $MCH_3CHNH_2COO^+$:

$$pK^0_1 = a' + \rho. \qquad (IV.2)$$

Substituting the value of ρ from this equation in the equation considered above, we have:

$$pK = a'' + nb \cdot pK^0_1, \qquad (IV.3)$$

where $a''=a\,(1-nba')$; pK^0_1 denotes the instability constant exponents for the corresponding alaninates, equal to 2.0 for Mg^{2+}, 3.0 for Mn^{2+}, 4.0 for Fe^{2+}, 4.8 for Co^{2+}, 6.0 for Ni^{2+}, 8.5 for Cu^{2+} and 5.2 for Zn^{2+}. Any ligand may be chosen as standard. In this case the alaninate ion was chosen since the instability constants for these complexes are most reliable.

TABLE 8

The constants in equation $pK = a + nbp$ for complexes formed by doubly-charged ions with the same volume (radius 0.8—0.9Å)

Ligand	No. of ligands	a	Relative polarizability
SO_4^{2-}	1	2.35	0.00
$P_2O_9^{3-}$	1	3.4	0.00
$P_4O_{12}^{4-}$	1	5.2	0.00
$S_2O_3^{2-}$	1	1.8	0.06
$CH_3CO_2^-$	1	0.7	0.18
$(CH_3CO_2)_2^{2-}$	1	1.8	0.18
$CH_2(CO_2)_2^{2-}$	1	2.0	0.41
$C_2O_4^{2-}$ CHO	1	2.5	0.47
$C_6H_3 - SO_3^-$ / O- CHO	1	0.0	0.62
C_6H_4 O- / OH-	1	1.5	0.67
$H_2NCH_2CONHCH_2COO^-$	1	1.4	0.62
	1	-0.2	0.46
(glycyl-glycine)	2	-2.0	0.80

Ligand	No. of ligands	a	Relative polarizability
8-Hydroxyquinolinate (oxinate)	1	5.0	0.96
$(OIIC_2H_4)_2NCH_2CO_2^-$	1	0.4	0.96
$H_2NCH_2COO^-$	1	0.9	0.90
$H_2NCH_2COO^-$	2	0.4	0.90
$CH_3CHNH_2COO^-$ $CH_2CO_2^-$	2	-1.0	1.00
$HN\diagdown C_2H_4CO_2^-$	1	0.5	1.12
$HN(C_2H_4CO_2)_2^{2-}$	1	-0.8	1.19
$HN(CH_2CO_2)_2^{2-}$	1	2.2	0.97
$HOC_2H_4N(CH_2CO_2)_2^{2-}$	1	1.1	1.44
NH_3	1	-1.0	0.61
NH_3 / $C_2H_4(NH_2)_2$ / $CH_3C_2H_3(NH_2)_2$	2 / 1 / 1	-2.2 / -1.7 / -1.7	0.59 / 1.44 / 1.44
Diethylene-triamine	1	-3.0	2.3

Equation (IV.3) has been checked for 27 series of complex compounds. It has been shown to be applicable only in those cases where not more than four co-ordination positions are occupied in the inner co-ordination sphere. It cannot therefore be applied to complex ammines of the type $M(NH_3)_5^{2+}$ and $M(NH_3)_6^{2+}$; to complex compounds of ethylenediamine of the type $M(en)_3^{2+}$; to ethylenediaminetetraacetate complexes, or to other similar compounds. In the cases where equation (IV.3) is applicable, the mean deviation in the pK values amounts to $\pm$ 0.15.

Equation (IV.3) may be used to check available data and to calculate unknown instability constants by interpolation and extrapolation. Table 8 gives the constants of equation (IV.3) for 27 series of complex compounds, calculated from 128 instability constant values.

The relative polarizability of a ligand is determined first of all by the nature of the atoms directly linked to the central atom: if the ligands are linked via oxygen, the value of b does not exceed 0.7; for ligands joined to the central atom via nitrogen, the value of b is always considerably larger. The presence of double bonds in the ligand ion always leads to an increase in the value of the relative polarizability.

The linear relationship which has been shown to exist between the instability constant exponents for complexes formed by ions of the same volume and charge may be used to predict the possible existence of new complexes and to estimate their stabilities. If complexes of a given type are shown to exist in solution for magnesium, copper and one of the intermediate members of the series, and if a linear relationship is established between the pK values, then complex compounds of the same type should also exist for the other four ions, and the instability constant exponents for these may be estimated by interpolation.

A physical interpretation of these regular features which have been shown to exist for the changes in the stability of complex compounds formed by ions with d-electrons may be given from the viewpoint of the crystal field theory [3, 4, 5]. The apparent deviations can also be explained from this standpoint.

The polarizing action of the central ion depends on its nuclear charge (atomic number), its radius, and the screening effect of the electrons present:

$$\rho = A\frac{Z-S}{r},\qquad\qquad\text{(IV.4)}$$

where A is a proportionality coefficient,

  Z is the atomic number,

  S is the screening constant, and

  r is the ionic radius.

The polarizing action of a spherical ion is the same in all directions; it remains the same for all ions in vacuo.

The main physical idea of the crystal field theory is that the electrons of the central ion avoid those regions where the negatively charged ligands (or the negative poles of molecular dipole ligands) create strong electrical fields.

If the ligands are situated at the apices of an octahedron, the degenerate d-level of the central ion splits into two new levels, the lower a triplet and the upper a doublet. The axes of the electron clouds of the upper level are directed towards the apices of the octahedron, while those of the lower level lie between these directions. The electrons of the lower level are denoted by d_ϵ, those of the higher level by d_γ.

An increase in the number of d_ϵ electrons increases the screening effect in directions perpendicular to the centres of the faces of the octahedron, and has no significant influence on the interaction between the central atom and the ligands. A decrease in the number of d_γ-electrons leads to a sharp decrease in the screening effect in directions pointing towards the apices of the octahedron, and as a result leads to a sharp increase in the polarizing action of the central ion in these directions. The greater the number of d_γ electrons, the weaker the polarizing action of the central ion, the weaker the chemical bond with ligands, and the smaller the probability of σ-bond formation.

Table 9 shows the distribution of electrons in the energy levels. In the case of octahedral complexes of the type MA_6, the stability in the series Ti^{2+}, V^{2+}, Cr^{2+}, Mn^{2+}, Fe^{2+}, Co^{2+}, Ni^{2+}, Cu^{2+}, Zn^{2+} should decrease at Cr^{2+} (the appearance of the first d_γ electron), Mn^{2+} (appearance of the second

d_τ electron), Cu^{2+} (appearance of the third d_τ electron) and Zn^{2+} (appearance of the fourth d_τ electron). The order of the change in stability of these complexes should be as follows:

$$Ti^{2+} < V^{2+} > Cr^{2+} > Mn^{2+} < Fe^{2+} < Co^{2+} < Ni^{2+} > Cu^{2+} > Zn^{2+}$$
$$Ti^{3+} < V^{3+} < Cr^{3+} > Mn^{3+} > Fe^{3+} < Co^{3+} < Ni^{3+} < Cu^{3+} > Ga^{3+}.$$

The increase in the stability of complex compounds from Mn^{2+} to Ni^{2+} and from Fe^{3+} to Cu^{3+} is related to the continuous increase in the polarizing action, resulting from the increase in the atomic number of the element. Experimental data confirm this order of change in the stability of the complex compounds.

A decrease in the symmetry of the electrical field created by the ligands leads to a further splitting of the energy levels and to the appearance of new sub-levels. Thus, when an axial field supplementary to the octahedral field is present, the upper d_τ level splits into two sub-levels: $d_{\tau'}$ and $d_{\tau'}$. The electron cloud of one of these ($d_{\tau'}$) is directed along a line joining opposite apices of the octahedron, while that of the second ($d_{\tau''}$) is directed in a plane perpendicular to this line, passing through the other four apices of the octahedron. The distribution of electrons in the various sub-levels when this axial field is present is given in the same table.

If the ligands occupy not more than four co-ordination positions lying in the same plane, then the screening effect is determined by the d_τ'' electrons, so that the following order of change in the stability of the complex compounds should be observed:

$$Ti^{2+} < V^{2+} < Cr^{2+} > Mn^{2+} < Fe^{2+} < Co^{2+} < Ni^{2+} < Cu^{2+} > Zn^{2+}$$
$$Ti^{3+} < V^{3+} < Cr^{3+} < Mn^{3+} > Fe^{3+} < Co^{3+} < Ni^{3+} < Cu^{3+} > Ga^{3+}$$

The decrease in the stability in the case of Mn^{2+} and Zn^{2+} complexes is explained by the appearance of the first and second screening $d_{\tau''}$ electrons respectively. This order of change in the stability of complexes of manganese, iron, cobalt, nickel, copper and zinc is observed in a large number of cases. In non-Soviet literature this order has been described as the Irving-Williams series. Yatsimirskii and Fed-

TABLE 9

Distribution of electrons within the energy levels in an octahedral field

Ions	No. of d_ε electrons	No. of d_γ electrons	Distribution of electrons under the influence of an axial field	
			No. of $d_{\gamma'}$ electrons	No. of $d_{\gamma''}$ electrons
Ti^{3+}	1	—	—	—
$Ti^{2+}V^{3+}$	2	—	—	—
$V^{2+}Cr^{3+}$	3	—	—	—
$Cr^{2+}Mn^{3+}$	3	1	1	—
$Mn^{2+}Fe^{3+}$	3	2	1	1
$Fe^{2+}Co^{3+}$	4	2	1	1
$Co^{2+}Ni^{3+}$	5	2	1	1
$Ni^{2+}Cu^{3+}$	6	2	1	1
Cu^{2+}—	6	3	2	1
$Zn^{2+}Ga^{3+}$	6	4	2	2

orova [6] have established the position of Cr^{2+} in this series; the complexes formed by the Cr^{2+} ion are much more stable than the corresponding manganese complexes.

It was shown above that a different order of change in the stability of complex compounds may exist when all six co-ordination positions are filled, or when five ligands are present.

An exactly analogous treatment is applicable to complex compounds with ligands situated at the apices of a tetrahedron. In the case of a tetrahedral field, the levels of lowest energy are d_γ levels and those of highest energy d_ε levels.

The axes of the electron clouds of the d_ε level are directed towards the apices of the tetrahedron; increase in the number of d_ε electrons leads to an intensification of the screening effect and a weakening of the bond between the central ion and the ligands. Table 10 shows the distribution of the electrons of the central ions when a tetrahedral field is present.

As may be seen from the table, in this case the maximum stability should be shown by tetrahedral complexes of Co^{2+}.

TABLE 10

Distribution of electrons within the
energy levels in a tetrahedral field

Ions	Number of d_γ-electrons	Number of d_ε-electrons
Mn^{2+}	2	3
Fe^{2+}	3	3
Co^{2+}	4	3
Ni^{2+}	4	4
Cu^{2+}	4	5
Zn^{2+}	4	6

The experimental data confirm this - the complex ions
$CoCl_4^{2-}$, $Co(CNS)_4^{2-}$ are particularly stable.

All the foregoing discussion relates to complexes with
ligands which are characterized by comparatively weak fields.
Increase in the field strength may lead to a decrease in the
number of screening electrons as a result of pairing of
d-electrons in the lower energy levels.

The change in the stability of complex compounds formed by
the series of ions in the middle of the fourth period of
Mendeleyev's periodic system is of a more complicated character.   In this case both maximum and minimum stabilities are
observed.   The position of the maximum stabilities may vary
with change in the spatial configuration of the complex ions
and in the number of co-ordination positions occupied.

Studies carried out in recent years show that the formation
of a dative π-bond, as a result of interaction between filled
d-orbitals in the central atom and vacant p- or d-orbitals in
the valency clouds of the ligand atoms, leads to an increase
in the stability of the complex compound.   The possible
formation of such bonds in complex compounds was first pointed
out by Pauling [7].   Syrkin and Dyatkina [8, 9] subsequently
developed this hypothesis, indicated the extensive occurrence
of π-bonds in complex compounds, and revealed new possibili-
ties for the formation of such bonds $(d_\pi - d_\pi$ bonds$)$.

At least three conditions must be observed if dati same bonds are to be formed: 1) the presence of a considdepend number (more than three) of free d-electrons in the central atom; 2) the presence, in the outer cloud of the same atom, of vacant p- or d-orbitals, by means of which the bonds between the ligand and central atom may be formed. The bonds formed may be $d_\pi - p_\pi$ or $d_\pi - d_\pi$- bonds, depending on whether p- or d-orbitals are vacant; 3) the formation of a sufficiently stable σ-bond between the central atom and one of the ligand atoms.

For the formation of a sufficiently stable σ-bond (the third condition), the central ion must have a high electron affinity. The electron affinity of a central ion M^{n+} is numerically equal to the nth ionization potential of the atom M, i.e., I_n. Another necessary condition for the formation of a dative σ-bond is a high mobility for the electrons of the ligand atoms, which is related to the magnitude of the polarizability of the ligands or their ionization potentials (the ionization potential of a ligand anion is numerically equal to the electron affinity of the electrically neutral atom).

The strength of the π-bonds depends on the mobility of the free d-electrons of the central atom. As a first approximation, the mobility of the d-electrons may be defined by the magnitude of the ionization potential of the ion M^{n+} (I_{n+1}) . The lower the value of I_{n+1}, the stronger the corresponding π-bond.

Since the formation of a dative π-bond leads to an increase in the effective positive charge on the central ion as a result of the partial transfer of d-electrons from the central atom to the ligands, the strength of the dative π-bonds decreases with increase in the charge on the central atom. Increase in the charge in a series of ions of the same structure leads to a sharp increase in the ionization potentials I_{n+1}, which should also lead to a decrease in the strength of the corresponding π-bond.

The strength of a π-bond should thus be greater, the greater the electron affinity of the cation, the lower its ionization potential and charge, and the lower the ionization potential of the ligand.

In the light of what has been said above, it can be seen

that the following ions are capable of forming dative π-bonds:
Cu^+, Cr^{2+}, Mn^{2+}, Fe^{2+}, Co^{2+}, Ni^{2+}, Cu^{2+}, Zn^{2+}; the same elements with
oxidation number $+3$ (except chromium) and Ga^{3+}, together with
the elements in the corresponding groups of the fifth and
sixth periods. Strong π-bonds cannot be formed by ions with
an inert gas type of electronic structure $(Li^+, Na^+, K^+,$
$Rb^+, Cs^+, Be^{2+}, Mg^{2+}, Ca^{2+}, Sr^{2+}, Ba^{2+}, Ra^{2+}, Al^{3+}, Sc^{3+}$, the lanthanide
ions, Ti^{4+} Zr^{4+}, Hf^{4+}, $Th^{4+})$ or by $In^+, Sn^{2+}, Sb^{3+}, Tl^+, Pb^{2+}$ and Bi^{3+}.

Dative π-bonds are formed by chloride, bromide and iodide
ions (but not by fluoride), by ligands directly linked to
the central atom via sulphur, selenium, phosphorus or arsenic
atoms, and by ligands in which multiple bonds are present
(cyanide, benzene, carbon monoxide, ethylene, etc.).

Consideration of the possibility of dative π-bond formation enables us to explain a number of apparent anomalies in
the changes in the stability of complex compounds.

It becomes easier, for example, to understand the decrease
in stability of cyanide, iodide, bromide, thiourea, thiosulphate and certain other complexes in the series Cu^+, Zn^{2+},
Ga^{3+} and Ag^+, Cd^{2+}, In^{3+} . This is related to the fact that
the strength of the π-bonds decreases with increase in the
charge of the central ion. The stability of complex compounds with ligands which are not capable of forming π-bonds
(for example, aliphatic amines) increases along the above
series. Ligands which have a marked tendency to form strong
π-bonds (cyanide, iodide, thiourea) stabilize the lower oxidation states, whereas ligands which do not form such bonds,
as a rule, stabilize the higher oxidation states.

The stability of the halide complexes of metals which do
not tend to form strong π-bonds decreases from the fluoride
to the iodide complexes. In the case of elements which have
a marked tendency to form dative π-bonds (mercury, platinum,
copper, silver, palladium, cadmium, etc.), however, the re-
verse order of change in stability is observed. This is
connected with the fact that the fluoride ions form no π-
bonds, while the tendency to form such bonds increases in the
series chloride-bromide-iodide.

The sharp decrease in the tendency to undergo complex-
formation, which is observed on passing from ions with 18

electrons in the outer shell to ions with 18 + 2 electrons
(Au^+—Tl^+, Hg^{2+}—Pb^{2+}, Tl^{3+}— Bi^{3+}) may be explained by the
impossibility of formation of π-bonds in the latter cases
(Tl^+, Pb^{2+}, Bi^{3+}).   For the same reason these ions do not form
stable cyanide complexes.   The thiocyanate complexes formed
by these ions are less stable than the chloride, bromide and
iodide complexes, whereas the thiocyanate complexes of ions
with d-electrons in the outer shell are of approximately the
same stability as the bromide complexes.   The stability of
the thiocyanate complexes is apparently greatly increased by
the formation of dative π-bonds.

Since the strength of the π-bonds increases with increase
in the number of d-electrons, the stability of compounds of
the type MCl^+ increases regularly in the series Mn^{2+}, Fe^{2+},
Co^{2+}, Ni^{2+}, Zn^{2+} .

The possibility of dative π-bond formation explains the
existence of complex compounds formed by ions with a large
number of d-electrons in the outer shell (Cu^+, Ag^+, Hg^{2+} etc.)
with aromatic hydrocarbons and a number of other molecules
with multiple bonds.

The nature of the ligand influences the stability of a
complex compound to the same extent as does the nature of the
central atom.   From what has been said above, we may accept
the following as being the most important ligand characteristics determining the stability of the corresponding complex
compounds.

1) Electrostatic characteristics: the charge and radius
(for ions) or the dipole moment (for molecules and ions).
The greater the electrostatic characteristics (maximum charge
and dipole moment, minimum radius), the more stable the complex compounds in question.   At the same time, as indicated
above, the formation of a stable complex requires a certain
correspondence between the electrostatic characteristics of
the central ion and the ligand.

2) The tendency of the ligand to form covalent σ-bonds.
This tendency is determined by the mobility of the outer
electrons and is therefore dependent on the ionization potential of the ligand ion (electron affinity of the atom), on
the polarizability of the ligand, and hence on the nature of
the atom by means of which the ligand is linked to the central

atom (increasing in the series O, N, S, Se etc.). As in the previous case, complex-formation requires a certain correspondence between the central atom and the ligand: the central atom should also have a tendency to form dative σ-bonds.

3) The possibility of dative π-bond formation by the ligand. This possibility arises when the valency cloud of the ligand atom contains vacant p-orbitals (ligands with multiple bonds - cyanide, aromatic compounds, etc.) or vacant d-orbitals (chlorine, bromine, iodine, sulphur, selenium, phosphorus, arsenic, etc.).

4) The number of co-ordination positions which may be occupied by one ligand particle. The greater the number of co-ordination positions occupied by a given ligand, the more stable the complex produced, since this leads to a more complete dehydration of the central ion, which in turn is related to a considerable positive entropy change.

The relationship between the stepwise instability constants is extremely important. According to Babko [10], the stepwise instability constant exponents (pk_n) are determined by the magnitude of the first instability constant exponent (pk_1) and a factor a_{n-1}, dependent on the value of the charges on M and A in the complex particle MA_n:

$$pk_n = a_{n-1} \cdot pk_1. \qquad (IV.5)$$

This equation is not applicable to complex compounds with neutral ligands; in this case Bjerrum's equation [11] is often employed:

$$pk_n = pk_1 - \log \frac{N(n-1)}{N-n}, \qquad (IV.6)$$

where N is the limiting number of ligands which may be attached to the central atom.

Recently, van Panthaleon van Eck [12] has put forward an empirical equation for the stepwise instability constant exponents:

$$pk_n = pk_1 - 2\lambda(n-1), \qquad (IV.7)$$

where λ is a constant.

Analysis of the experimental data shows good agreement with this empirical equation. In particular, good agreement with experiment is observed in the case of fluoride and certain bromide complexes and for many complex compounds with ammonia, imidazole and hydrazine. The van Panthaleon equation is apparently very accurate in the case where the ligands lie at the apices of an octahedron.

The change in the magnitude of the stepwise constants with change in the number of attached ligands should undoubtedly be determined by the structural features of the complex compounds in question, so that there cannot be a single, universal equation which does not take account of these stereochemical factors.

In the stepwise formation of a complex ion with ligand anions situated at the apices of a tetrahedron, for example, the mutual repulsion of the anions in the particle MA_2 will be greater than the analogous repulsion in an octahedral complex ion, where all three MA_2 particles lie on the same straight line. This means that in the case of tetrahedral co-ordination, the second instability constants have values which are too high, so that the pk_2 values do not lie in the series. This phenomenon is observed in the acido-complexes of indium and cadmium.

The particular type of substitution in the complex compounds of mercury [13] leads to a particular type of change in the order of the instability constants of these compounds.

In order to estimate the relationship between the stepwise instability constants, therefore, it is necessary to know not only the formula of the complex compound but also its stereochemical characteristics.

REFERENCES

1. A. A. GRINBERG and K. B. YATSIMIRSKII, Izv. Akad. Nauk SSSR, Otd. khim. nauk, No. 2, 211 (1952).
2. K. B. YATSIMIRSKII, Zh. obshch. khim., 24, 1498 (1954).
3. L. ORGEL, J. Chem. Soc., 4756 (1952).
4. J. BJERRUM and C. K. JORGENSEN, Rec. trav. chim., 75, 658 (1956).
5. K. B. YATSIMIRSKII, Zh. neorg. khim., 1, 2451 (1956).
6. K. B. YATSIMIRSKII and T. I. FEDOROVA, ibid., 1, 2310 (1956).
7. L. P. PAULING, The Nature of the Chemical Bond (Priroda khimicheskoi svyazi), (Translation into Russian), Moscow (1947).

8. Ya. K. SYRKIN and M. E. DYATKINA, Zh. obshch. khim.,
    16, 345 (1946).
9. Ya. K. SYRKIN, Izv. Akad. Nauk SSSR, Otd. khim. nauk,
    No. 1, 69 (1948).
10. A. K. BABKO, The Physico-Chemical Analysis of Complex
    Compounds in Solution (Fiziko-khimicheskii analiz
    kompleksnykh soyedinenii v rastvorakh), Kiev (1955).
11. J. BJERRUM, Metal Ammine Formation in Aqueous Solutions,
    Copenhagen (1941).
12. C. L. van PANTHALEON van ECK, Rec. trav. chim., 72, 529
    (1953).
13. C. L. van PANTHALEON van ECK, H. B. M. WALTERS and
    W. J. JASPERS, ibid., 75, 796 (1956).

Tables
of Instability Constants
of Complex Compounds

INTRODUCTORY NOTE TO TABLES

The Tables give data on the instability constants of 1381 complex particles, the majority of which contain only one type of ligand. The data have been obtained by searching the literature up to 1954, and in some cases up to 1955-1956.

Only a few instability constants of polynuclear complexes and complexes with mixed ligand types are given here.

1. The Arrangement of the Material

The instability constants are arranged in groups relating to complexes with the same ligands (for example, ammines, bromide complexes, chloride complexes). These groups of complexes are divided as follows:

1) complexes with inorganic ligands;

2) complexes with organic ligands.

In the first section (complexes with inorganic ligands), the groups of complexes are arranged for the most part in alphabetical order of ligand names*.

In the second section (complexes with organic ligands), the specific features of organic nomenclature and a number of other factors have led us to adopt the following sub-divisions:

1) complexes with amines;

2) complexes with organic acid anions;

---

* Presented in this translation in similarly approximate English alphabetical order of English names of ligands - Translator.

3) complexes with amino-acids;

4) complexes with diketones and aldehydes;

5) complexes with other organic ligands.

In each of these sub-divisions, the groups of complexes are arranged for the most part in alphabetical order of the ligand names. A list of all the ligands is given separately (pp. 91-93); for convenience in using the Tables, the order of arrangement of the ligands in this list is the same as in the Tables, a consolidated index being also given at p.217.

In all the Tables, the complexes within a given group with the same ligand are arranged in alphabetical order of the chemical symbols of the central ion (Ag, Al, Au, etc.)

2. Table Contents and Abbreviations Used

After the heading, which gives the name of the group of complexes (for example, acetate complexes, ethylenediamine-tetraacetate complexes), there is given the chemical formula of the ligand, for example,

$$CH_3COO - , \quad \begin{array}{c} H_2C-N \Big\langle \begin{array}{c} CH_2COO- \\ CH_2COO- \end{array} \\ | \\ H_2C-N \Big\langle \begin{array}{c} CH_2COO- \\ CH_2COO- \end{array} \end{array}$$

Where the ligand formula is cumbersome, there is given, in addition to the chemical formula, an arbitrary abbreviation for the ligand (for example, for

$$\begin{array}{c} H_2C-N \Big\langle \begin{array}{c} CH_2COO- \\ CH_2COO- \end{array} \\ | \\ H_2C-N \Big\langle \begin{array}{c} CH_2COO- \\ CH_2COO- \end{array} \end{array} , \ Edta^{4-}),$$

which is then used throughout the group of complexes in question.

The first column of the Table ("Complex Ion") gives the

formula of the complex compound [for example, $Cu(CH_3COO)_2$ or $NiEdta^{2-}$].

The second column ("Temperature") gives the temperature, in °C, at which the determination of the instability constant was carried out. In some cases the indication "room" (room temperature) is given, since the original sources of the data are not more specific. In cases where the temperature is not indicated at all the presumption seems to be that room temperature is meant.

The third column ("Ionic Strength") gives the value of the ionic strength, which is equal to half the sum of the products of the ion concentrations and the square of their charges. The absence of a figure in this column indicates that the ionic strength is not given in the original. In these cases the values of the instability constants evidently relate to solutions with the finally reached ionic strengths (i.e., without extrapolation to zero ionic strength).

The fourth column ("Method") indicates the method used for the determination of the instability constants. In this case the following abbreviations have been used:

1. Analysis - Analytical determination of the concentrations.

2. Biol. - Biological method.

3. Ind. pH meas. (pH-color.) - Indicator measurement of pH.

4. Ion exch. - Ion exchange method.

5. Kin. - Kinetic method.

6. Cryosc. - Cryoscopic method.

7. Polarog. - Polarographic method.

8. Potent. - Potentiometric method (pH-potent.).

9. Distrib. - Distribution between immiscible liquids.

10. Solub. - Solubility method.

11. Calc. - By calculation.

12. Spectr. - Spectrophotometric method.

13. Thermodyn. - Thermodynamic method.

14. El. cond. - Electrical conductivity method.

The fifth column (k) gives the values of the successive instability constants k, i.e., the equilibrium constants for the reaction:

$$MA_i \rightleftarrows MA_{i-1} + A,$$

i.e.

$$k = \frac{[MA_{i-1}][A]}{[MA_i]}.$$

The sixth column (pk) gives the successive instability constant exponents pk, i.e., the reciprocal of the logarithm of the constants:

$$pk = -\log k.$$

The seventh column (K) gives the values of the overall instability constants, i.e., the equilibrium constants for the reaction:

$$MA_i \rightleftarrows M + iA,$$

$$K = \frac{[M][A]^i}{[MA_i]}.$$

In the cases where $i = 1$, the values for the successive and overall instability constants coincide.

The eighth column (pK) gives the overall instability constant exponents pK, i.e., the reciprocal of the logarithm of the instability constants:

$$pK = -\log K.$$

In the cases where different values for the instability constant of a particular complex are given by different authors, we have chosen the value which appears to us to be the most reliable: in a number of cases, however, the instability constant values determined by different authors show good agreement.

The values of k or pk, K or pK given in the Tables have the same number of significant figures as are given in the original. Since an estimate of the accuracy of the determination of the constants in a number of works is not given, the number of significant figures does not always indicate the actual accuracy of the determination.

Where instability constant values are given in the Tables for two ionic strengths, the less reliable figure is given in brackets.

The ninth and tenth columns ("References") give the literature references. The ninth column ("Principal reference") 'gives the literature source reporting the determination of the instability constant value given in the Table. The tenth column ("Supplementary references") gives sources which report numerical values for the instability constants determined for the most part under different conditions of temperature or ionic strength, together with figures which, for a variety of reasons, have not been given in our Tables.

3. LIST OF LIGANDS

I. INORGANIC LIGANDS

1.	Ammonia	13.	Nitrate
2.	Bromate	14.	Selenocyanate
3.	Hydrazine	15.	Phosphate
4.	Bromide	16.	Pyrophosphate
5.	Chlorate	17.	Perchlorate
6.	Iodate	18.	Sulphate
7.	Chloride	19.	Sulphite
8.	Hydroperoxide	20.	Tetrametaphosphate
9.	Cyanide	21.	Thiocyanate
10.	Fluoride	22.	Thiosulphate
11.	Hydroxyl	23.	Thiourea
12.	Iodide	24.	Trimetaphosphate

II. ORGANIC LIGANDS

1. Amines

1.	Diethylenetriamine	7.	Pyridine
2.	Dipyridyl	8.	1,2,3-Triaminopropane
3.	Ethylenediamine	9.	Methylamine
4.	Imidazole	10.	Triaminotriethylamine
5.	Phenanthroline	11.	Trimethylenediamine
6.	Propylenediamine	12.	Triethylenetetramine

2. Organic acid anions

1.	Acetate	10.	Malate
2.	Butyrate	11.	Nitroacetate
3.	Citrate	12.	Oxalate
4.	Glycerate	13.	Oxalacetate
5.	Glycollate	14.	Phthalate
6.	Gluconate	15.	Salicylate
7.	Kojic acid anion	16.	Succinate
8.	Lactate	17.	Tartrate
9.	Malonate	18.	Propionate

3. Aminoacids

1. Alanine
2. 2-Sulphoanilinediacetic acid
3. Aminobarbituric-N,N-diacetic acid
4. Asparagine
5. N-Hydroxyethylethylenediaminetriacetic acid
6. Aspartic acid
7. 1,2-Diaminocyclohexanetetraacetic acid
8. Ethylenediaminetetraacetic acid
9. Glycine
10. N,N-Dihydroxyethylglycine
11. Glycyl-glycine
12. β-Hydroxyethyliminodiacetic acid
13. Iminodiacetic acid
14. Iminodipropionic acid
15. Iminopropionicacetic acid
16. Nitrilodiaceticpropionic acid
17. Nitrilodipropionicacetic acid
18. Nitrilotripropionic acid
19. Nitrilotriacetic acid
20. Trimethylenediaminetetraacetic acid

4. Diketones and Aldehydes

1.	Acetylacetone	6.	Salicylaldehyde
2.	β-Methyltropolone	7.	α-Isopropyltropolone
3.	Sulphosalicylaldehyde	8.	β-Isopropyltropolone
4.	Thenoyltrifluoroacetone	9.	α-Methyltropolone
5.	Tropolone		

5. Other Organic Ligands

1. o-Aminophenol
2. Eriochrome Black A
3. Eriochrome Black T
4. Eriochrome Blue Black B
5. Eriochrome Blue Black R
6. 8-Hydroxy-2,4-dimethylquinazoline
7. 8-Hydroxy-4-methyl-2-phenylquinazoline
8. 8-Hydroxy-5-methylquinoline
9. 8-Hydroxy-2-methylquinoline
10. 8-Hydroxy-6-methylquinoline
11. 8-Hydroxy-7-methylquinoline
12. 8-Hydroxy-4-methylcinnoline
13. 8-Hydroxyquinazoline
14. 5-Hydroxyquinoxaline
15. 8-Hydroxycinnoline
16. 8-Hydroxyquinoline

I. COMPLEXES WITH INORGANIC LIGANDS

Ammonia Complexes

Complex ion	Temperature °C	Ionic strength	Method	k	pk	K	pK	References principal	supplementary
$AgNH_3^+$	30	0.5—5.0	pH-potent.	$6.30 \cdot 10^{-4}$	3.20	$6.30 \cdot 10^{-4}$	3.20	[1]	[5]
$Ag(NH_3)_2^+$	30	0.5—5.0	"	$1.48 \cdot 10^{-4}$	3.83	$9.31 \cdot 10^{-8}$	7.03	[1]	[3,6]
$CdNH_3^{2+}$	30	0.5—5.0	"	$2.24 \cdot 10^{-3}$	2.65	$2.24 \cdot 10^{-3}$	2.65	[1]	[7,8]
$Cd(NH_3)_2^{2+}$	30	0.5—5.0	"	$7.94 \cdot 10^{-3}$	2.10	$1.78 \cdot 10^{-5}$	4.75	[1]	[7,8]
$Cd(NH_3)_3^{2+}$	30	0.5—5.0	"	$3.63 \cdot 10^{-2}$	1.44	$6.46 \cdot 10^{-7}$	6.19	[1]	[7,8,9]
$Cd(NH_3)_4^{2+}$	30	0.5—5.0	"	$1.17 \cdot 10^{-1}$	0.93	$7.56 \cdot 10^{-8}$	7.12	[1]	[8,10]
$Cd(NH_3)_5^{2+}$	30	0.5—5.0	"	2.1	−0.32	$1.6 \cdot 10^{-7}$	6.80	[1]	—
$Cd(NH_3)_6^{2+}$	30	0.5—5.0	"	46.0	−1.66	$7.3 \cdot 10^{-6}$	5.14	[1]	—
$CoNH_3^{2+}$	30	0.5—5.0	"	$7.75 \cdot 10^{-3}$	2.11	$7.75 \cdot 10^{-3}$	2.11	[1]	—
$Co(NH_3)_2^{2+}$	30	0.5—5.0	"	$2.34 \cdot 10^{-2}$	1.63	$1.81 \cdot 10^{-4}$	3.74	[1]	—
$Co(NH_3)_3^{2+}$	30	0.5—5.0	"	$8.90 \cdot 10^{-2}$	1.05	$1.62 \cdot 10^{-5}$	4.79	[1]	—
$Co(NH_3)_4^{2+}$	30	0.5—5.0	"	$1.73 \cdot 10^{-1}$	0.76	$2.80 \cdot 10^{-6}$	5.55	[1]	—
$Co(NH_3)_5^{2+}$	30	0.5—5.0	"	$6.6 \cdot 10^{-1}$	0.18	$1.85 \cdot 10^{-6}$	5.73	[1]	—
$Co(NH_3)_6^{2+}$	30	0.5—5.0	"	4.2	−0.62	$7.75 \cdot 10^{-6}$	5.11	[1]	—
$Co(NH_3)_6^{3+}$	30	1.0—2.0	"	—	—	$3.1 \cdot 10^{-33}$	32.51	[1]	—
$CuNH_3^{2+}$	30	0.5—5.0	"	$7.10 \cdot 10^{-5}$	4.15	$7.10 \cdot 10^{-5}$	4.15	[1]	—
$Cu(NH_3)_2^{2+}$	30	0.5—5.0	"	$3.16 \cdot 10^{-4}$	3.50	$2.24 \cdot 10^{-8}$	7.65	[1]	[11]
$Cu(NH_3)_3^{2+}$	30	0.5—5.0	"	$1.29 \cdot 10^{-3}$	2.89	$2.89 \cdot 10^{-11}$	10.54	[1]	—
$Cu(NH_3)_4^{2+}$	30	0.5—5.0	"	$7.40 \cdot 10^{-3}$	2.13	$2.14 \cdot 10^{-13}$	12.67	[1]	[11,12]
$Cu(NH_3)_5^{2+}$	30	—	"	3.0	−0.5	—	—	[2]	—

Ammonia complexes (contd.)

Complex ion	Temperature °C	Ionic strength	Method	h	pk	K	pK	References principal	References supplementary
$CuNH_3^+$	—	—	—	$6.6 \cdot 10^{-7}$	6.18	$6.6 \cdot 10^{-7}$	6.18	[3]	—
$Cu(NH_3)_2^+$	—	—	—	$2.04 \cdot 10^{-5}$	4.69	$1.35 \cdot 10^{-11}$	10.87	[3]	[12]
$FeNH_3^{2+}$	20—30	0.5—5.0	Calc.	$4.0 \cdot 10^{-2}$	1.4	$4.0 \cdot 10^{-2}$	1.4	[4]	—
$Fe(NH_3)_2^{2+}$	20—30	0 5—5 0	.	$1.6 \cdot 10^{-1}$	0.8	$6.4 \cdot 10^{-3}$	2.2	[4]	—
$HgNH_3^{2+}$	22	2.0	pH-potent	$1.6 \cdot 10^{-9}$	8.8	$1.6 \cdot 10^{-9}$	8.8	[1]	—
$Hg(NH_3)_2^{2+}$	22	2.0	.	$2.0 \cdot 10^{-9}$	8.7	$3.2 \cdot 10^{-18}$	17.5	[1]	—
$Hg(NH_3)_3^{2+}$	22	2.0	.	$1.0 \cdot 10^{-1}$	1.00	$3.2 \cdot 10^{-19}$	18.5	[1]	—
$Hg(NH_3)_4^{2+}$	22	2 0	.	$1.66 \cdot 10^{-1}$	0.78	$5.3 \cdot 10^{-20}$	19.28	[1]	—
$MgNH_3^{2+}$	22	2.0	.	$5.9 \cdot 10^{-1}$	0.23	$5.9 \cdot 10^{-1}$	0.23	[1]	—
$Mg(NH_3)_2^{2+}$	22	2.0	.	1.41	-0.15	$8.3 \cdot 10^{-1}$	0.08	[1]	—
$Mg(NH_3)_3^{2+}$	22	2 0	.	2.63	-0.42	2 8	-0.34	[1]	—
$Mg(NH_3)_4^{2+}$	22	2.0	.	5.0	-0.7	10.9	-1.04	[1]	—
$Mg(NH_3)_5^{2+}$	22	2.0	.	8.9	-0.95	97.0	-1.99	[1]	—
$Mg(NH_3)_6^{2+}$	22	2 0	.	20.0	-1.3	1930.0	-3.29	[1]	—
$MnNH_3^{2+}$	20—30	0.5—5.0	Calc.	$1.6 \cdot 10^{-1}$	0.8	$1.6 \cdot 10^{-1}$	0.8	[4]	—
$Mn(NH_3)_2^{2+}$	20—30	0.5—5.0	"	$3.2 \cdot 10^{-1}$	0.5	$5.0 \cdot 10^{-2}$	1.3	[4]	—
$NiNH_3^{2+}$	30	0.5—5.0	pH-potent	$1.62 \cdot 10^{-3}$	2.79	$1.62 \cdot 10^{-3}$	2.79	[1]	—
$Ni(NH_3)_2^{2+}$	30	0.5—5.0	.	$5.75 \cdot 10^{-3}$	2.24	$9.31 \cdot 10^{-6}$	5.03	[1]	—
$Ni(NH_3)_3^{2+}$	30	0.5—5.0	.	$1.86 \cdot 10^{-2}$	1.73	$1.73 \cdot 10^{-7}$	6.76	[1]	—
$Ni(NH_3)_4^{2+}$	30	0.5—5.0	.	$6.45 \cdot 10^{-2}$	1.19	$1.12 \cdot 10^{-8}$	7.95	[1]	—
$Ni(NH_3)_5^{2+}$	30	0.5—5.0	.	$1.78 \cdot 10^{-1}$	0.75	$2.00 \cdot 10^{-9}$	8.70	[1]	—
$Ni(NH_3)_6^{2+}$	30	0.5—5.0	.	$9.34 \cdot 10^{-1}$	0 03	$1.86 \cdot 10^{-9}$	8.73	[1]	—

Ammonia complexes (contd.)

Complex ion	Temperature °C	Ionic strength	Method	k	pk	K	pK	References	
								principal	supplementary
$TlNH_3^+$	16	—	Spectr.	8.3	-0.92	8.3	-0.92	[3]	—
$ZnNH_3^{2+}$	30	0.5—5.0	pH-potent	$4.26 \cdot 10^{-3}$	2.37	$4.26 \cdot 10^{-3}$	2.37	[1]	—
$Zn(NH_3)_2^{2+}$	30	0.5—5.0	.	$3.63 \cdot 10^{-3}$	2.44	$1.54 \cdot 10^{-5}$	4.81	[1]	[11]
$Zn(NH_3)_3^{2+}$	30	0.5—5.0	.	$3.16 \cdot 10^{-3}$	2.50	$4.87 \cdot 10^{-8}$	7.31	[1]	—
$Zn(NH_3)_4^{2+}$	30	0.5—5.0	.	$7.10 \cdot 10^{-3}$	2.15	$3.46 \cdot 10^{-10}$	9.46	[1]	[11]

REFERENCES

1. J. BJERRUM, Metal Ammine Formation in Aqueous Solutions, Copenhagen, 1941; cited in Chem. Abs., 35, 6527 (1941).

2. J. BJERRUM, Chem. Rev., 46, 381 (1950).

3. P. JOB, Ann. chim., [10] 9, 113; cited in Z. Bl., I, 2572 (1928).

4. K. B. YATSIMIRSKII, Zh. obshch. khim., 24, 1498 (1954).

5. W. C. VOSBURGH, McCLURE, J. Am. Chem. Soc., 65, 1060 (1943).

6. G. BODLANDER and R. FITTIG, Z. phys. Chem., 39, 597 (1902).

7. P. F. DERR and W. C. VOSBURGH, J. Am. Chem. Soc., 65, 2408 (1943).

8. C. G. SPIKE and R. W. PARRY, ibid., 75, 2726 (1953).

9. I. A. KORSHUNOV and L. V. LIPATOVA, Zh. obshch. khim., 21, 615 (1951).

10. A. G. STROMBERG and I. E. BYKOV, ibid., 19, 245 (1949).

11. C. G. SPIKE and R. W. PARRY, J. Am. Chem. Soc., 75, 3770 (1953).

12. M. v. STACKELBERG and H. v. FREYHOLD, Z. Elektrochem., 46, 120 (1940).

TABLES 97

Bromate Complexes

Complex ion	Temperature °C	Ionic strength	Method	k	pk	K	pK
$ThBrO_3^{3+}$	25	0.5	Distrib.	0.155	0.81	0.155	0.81
$Th(BrO_3)_2^{2+}$	25	0.5		0.79	0.10	0.123	0.91

REFERENCE

R. A. DAY and R. W. STOUGHTON, J. Am. Chem. Soc., 72, 5662
(1950).

Hydrazine Complexes

Complex ion	Temperature °C	Ionic strength	Method	k	pk	K	pK
$NiN_2H_4^{2+}$	20	0.5	pH-potent.	$1.75 \cdot 10^{-3}$	2.76	$1.75 \cdot 10^{-3}$	2.76
$Ni(N_2H_4)_2^{2+}$	20	0.5	"	$3.6 \cdot 10^{-3}$	2.44	$6.3 \cdot 10^{-6}$	5.20
$Ni(N_2H_4)_3^{2+}$	20	0.5	"	$7.1 \cdot 10^{-3}$	2.15	$4.5 \cdot 10^{-8}$	7.35
$Ni(N_2H_4)_4^{2+}$	20	0.5	"	$1.4 \cdot 10^{-2}$	1.85	$6.3 \cdot 10^{-10}$	9.20
$Ni(N_2H_4)_5^{2+}$	20	0.5	"	$2.8 \cdot 10^{-2}$	1.55	$1.9 \cdot 10^{-11}$	10.72
$Ni(N_2H_4)_6^{2+}$	20	0.5	"	$5.75 \cdot 10^{-2}$	1.24	$1.1 \cdot 10^{-12}$	11.96
$ZnN_2H_4^{2+}$	20	0.5	"	$4.0 \cdot 10^{-3}$	2.4	$4.0 \cdot 10^{-3}$	2.4
$Zn(N_2H_4)_2^{2+}$	20	0.5	"	$1.6 \cdot 10^{-2}$	1.8	$6.3 \cdot 10^{-5}$	4.2
$Zn(N_2H_4)_3^{2+}$	20	0.5	"	$5.0 \cdot 10^{-2}$	1.3	$3.26 \cdot 10^{-6}$	5.5
$Zn(N_2H_4)_4^{2+}$	20	0.5	"	$1.6 \cdot 10^{-1}$	0.8	$5.0 \cdot 10^{-7}$	6.3

REFERENCE

G. SCHWARZENBACH and A. ZOBRIST, Helv. chim. Acta., 35, 1291
(1952).

Bromide Complexes

Complex ion	Temperature °C	Ionic strength	Method	k	pk'	K	pK	References principal	References supplementary
Ag_2Br^+	Room	—	Solub.		—	$2 \cdot 10^{-10}$	9.70	[1]	—
$AgBr$	25	0.2	"	$7.1 \cdot 10^{-8}$	4.15	$7.1 \cdot 10^{-5}$	4.15	[2]	[23]
$AgBr_2^-$	25	0.2	"	$1.1 \cdot 10^{-8}$	2.96	$7.8 \cdot 10^{-8}$	7.11	[2]	[24]
$AgBr_3^{2-}$	25	0.2	"	$1.6 \cdot 10^{-2}$	1.79	$1.3 \cdot 10^{-9}$	8.90	[2]	[23,24,25,26]
$AgBr_4^{3-}$	25	0.2	Potent.	0.50	0.30	$6.3 \cdot 10^{-10}$	9.20	[2]	[27]
$AuBr_2$	60	3	"		—	$4 \cdot 10^{-13}$	12.4	[3]	—
$BiBr^{2+}$	18	—	Solub.	$5 \cdot 10^{-5}$	4.30	$5 \cdot 10^{-5}$	4.30	[4]	—
$BiBr_2^+$	25	1.0—2.0	"	$5.6 \cdot 10^{-2}$	1.25	$2.8 \cdot 10^{-6}$	5.55	[5]	[4]
$BiBr_3$	25	1.0—2.0	Potent.	0.48	0.32	$1.3 \cdot 10^{-6}$	5.89	[5]	[4]
$BiBr_4^-$	20	1.5—1.6	"		—	$1.5 \cdot 10^{-8}$	7.82	[5]	[4]
$BiBr_6^{3-}$	20	1.5—1.6	"		—	$2 \cdot 10^{-10}$	9.70	[6]	—
$CdBr^+$	25	3.0	"	$1.78 \cdot 10^{-2}$	1.75	$1.78 \cdot 10^{-2}$	1.75	[7]	[28,29]
$CdBr_2$	25	3.0	"	0.26	0.59	$4.5 \cdot 10^{-3}$	2.34	[7]	[28,29]
$CdBr_3^-$	25	3.0	"	0.105	0.98	$4.75 \cdot 10^{-4}$	3.32	[7]	[28,29,30]
$CdBr_4^{2-}$	25	3.0	Ion exch.	0.26	0.38	$2 \cdot 10^{-4}$	3.70	[7]	[28,29,20,15]
$CeBr^{2+}$	Room	0.0	Potent.	0.42	0.38	0.42	0.38	[8]	—
$CuBr_2$	18—20	0.02—0.5	Spectr.		—	$1.3 \cdot 10^{-6}$	5.89	[9]	—
$CuBr^+$	22	1.0	Potent.	0.5	0.30	0.5	0.30	[10]	—
$FeBr^{2+}$	26.7	1.0	"	2.0	-0.30	2.0	-0.30	[11]	[31,32]
$HgBr^+$	25	0.5	"	$0.89 \cdot 10^{-9}$	9.05	$0.89 \cdot 10^{-9}$	9.05	[12]	—
$HgBr_2$	25	0.5	"	$5.4 \cdot 10^{-9}$	8.28	$4.8 \cdot 10^{-18}$	17.32	[12]	[33]
$HgBr_3^-$	25	0.5	"	$3.8 \cdot 10^{-2}$	1.42	$1.82 \cdot 10^{-20}$	19.74	[12]	[33,34]
$HgBr_4^{2-}$	25	0.5	Ion exch.	$5.5 \cdot 10^{-2}$	1.26	$1.0 \cdot 10^{-21}$	21.00	[12]	[34]
$JnBr^{2+}$	25	1.0	"	$6.3 \cdot 10^{-2}$	1.20	$6.3 \cdot 10^{-2}$	1.20	[13]	[35]
$JnBr_2^+$	25	1.0	"	0.26	0.58	$1.7 \cdot 10^{-2}$	1.78	[13]	[35]
$JnBr_3$	25	1.0	"	0.20	0.70	$3.3 \cdot 10^{-3}$	2.48	[13]	[35]

Bromide complexes (contd.)

Complex ion	Temperature °C	Ionic strength	Method	k	p^k	K	pK	References principal	References supplementary
PbBr+	25	0	El. cond.	$3.3 \cdot 10^{-2}$	1.15	$3.3 \cdot 10^{-2}$	1.15	[14]	[36]
PbBr$_3^-$	25	0.0	Polarog.	—	—	$1.2 \cdot 10^{-2}$	1.92	[15]	—
PbBr$_4^{2-}$	25	0.0	Potent.	—	—	$1.0 \cdot 10^{-3}$	3.0	[15]	—
PdBr$_4^{2-}$	Room	—	"	—	—	$8.0 \cdot 10^{-14}$	13.1	[16]	—
PtBr$_4^{2-}$	.	—	.	—	—	$3.0 \cdot 10^{-21}$	20.5	[16[	[37]
SnBr+	25	3.0	.	$1.9 \cdot 10^{-1}$	0.73	$1.9 \cdot 10^{-1}$	0.73	[17]	—
SnBr$_2$	25	3.0	.	0.40	0.40	$7.2 \cdot 10^{-2}$	1.14	[17]	—
SnBr$_3^-$	25	3.0	.	0.62	0.21	$4.5 \cdot 10^{-2}$	1.35	[17]	—
TlBr^{2+}	25	0.1—0.2	.	$2 \cdot 10^{-10}$	9.7	$2 \cdot 10^{-10}$	9.7	[18]	—
		1.2			8.9		8.9	[19]	—
TlBr$_2^+$	25	0.1—0.2	.	$1.3 \cdot 10^{-7}$	6.9	$2.5 \cdot 10^{-17}$	16.6	[18]	—
		1.2			7.5		16.4	[19]	—
TlBr$_3$		0.1—0.2	.	$2.5 \cdot 10^{-5}$	4.6	$1.6 \cdot 10^{-22}$	21.2	[18]	—
		1.2			5.7		22.1	[19]	—
TlBr$_4^-$		0.1—0.2	.	$2 \cdot 10^{-3}$	2.7	$1.3 \cdot 10^{-24}$	23.9	[18]	—
		1.2			4.0		26.1	[19]	—
TlBr$_5^{2-}$		1.2			3.1		29.2	[19]	—
TlBr$_6^{3-}$		1.2			2.4		31.6	[19]	—
UBr^{3+}	20	—		0.67	0.18	0.67	0.18	[20]	—
UO$_2$Br+	20	—		2.00	−0.30	2.00	−0.30	[21]	—
ZnBr+	25	0.3		4.00	−0.30	4.0	−0.60	[22]	—

REFERENCES

1. K.B.YATSIMIRSKII, Dokl.Akad.Nauk SSSR, 77, 819 (1951).

2. E.BERNE and J.LEDEN, Z.Naturforsch., 8a, 719 (1953).

3. G. GRUBE and T. MORITA, Z.Elektrochem., 38, 117 (1932).

4. A.K. BABKO, Naukovi Zapiski KDU, 4, 81 (1939).

5. K.B.YATSIMIRSKII, Collected Articles on General Chemistry (Sbornik statei po obshchei khimii), Izd.Akad. Nauk SSSR, I, 97 (1953).
6. A.K.BABKO and A.M.GOLUB, ibid., I, 64.
7. J.LEDEN, Z.phys.Chem., 188, 160 (1941).
8. S.W.MAYER and S.D.SCHWARTZ, J.Am.Chem.Soc., 73, 222 (1951).
9. G.BODLEANDER and O.STORBECK, Z.anorg.Chem., 31, 458 (1902).
10. P.S.FARRINGTON, J.Am.Chem.Soc., 74, 966 (1952).
11. E.RABINOWITCH and W.STOCKMAYER, ibid., 64, 335 (1942).
12. O.BETHGE, J.JONEVALL-WESTOO and L.G.SILLEN, Chem.Abs., 43, 4545 (1949).
13. J.A.SCHUFFE and H.M.EILAND, J.Am.Chem.Soc., 76, 960 (1954).
14. G.H.NANCOLLAS, J.Chem.Soc., 1458 (1955).
15. A.M.VASIL'EV and V.I.PROUKHINA, Zh.anal.khim., 6, 218 (1951).
16. W.M.LATIMER, Okislitel'nye sostoyaniya elementov i ikh potensialy v vodnykh rastvorakh (The Oxidation States of the Elements and their Potentials in Aqueous Solution) (Translation into Russian), IL, Moscow (1954).
17. C.E.VANDERZEE, J.Am.Chem.Soc., 74, 4806 (1952).
18. R. BENOIT, Bull.Soc.chim.France, 518 (1949); Chem. Abs., 43, 8939 (1949).
19. D. PESCHANSKI and S. VALLADAS-DUBOIS, Compt.rend., 241, 1046 (1955).
20. S. AHRLAND and R. LARSSON, Acta Chem.Scand., 8, 137 (1954).
21. S. AHRLAND, ibid., 5, 1271 (1951).
22. L.G.SILLEN and B.LILJEQVIST, Chem.Abs., 40, 4588 (1946).
23. V.B.VOUK, J.KRATOHVIL and B.TEZAK, Arhiv Kem., 23, 200 (1951).
24. K.S.LYALIKOV and V.N.PISKUNOVA, Zh.fiz.khim., 28, 127 (1954).
25. W. ERBER, Z.anorg.Chem., 248, 32 (1941).
26. H.CHATEAU and J.POURADIER, Chem.Abs., 40, 8563 (1952).
27. G.BODLANDER and W.EBERLEIN, Z.anorg.Chem., 39, 197 (1904).
28. H.L.RILEY and V. GALLAFENT, J.Chem.Soc., 514 (1932).
29. L.ERIKSSON, Acta Chem. Scand., 7, 1146 (1953).
30. I.A.KORSHUNOV, N.I.MALYUTINA and O.M.BALABANOVA, Zh. obshch. khim., 21, 620 (1951).
31. R.NASANEN, Chem. Abs., 44, 10415 (1950).
32. H.L.RILEY and H.C.SMITH, J. Chem. Soc., 1448 (1934).
33. H. MORZE, Z.phys.Chem., 41, 709 (1902).
34. M.S.SHERILL, ibid., 43, 705 (1903); 47, 103 (1904).
35. N.SUNDEN, Svensk Kem. Tidsk., 66, 20 (1954); Chem.Abs., 48, 9252 (1954).
36. H.FROMHERZ, Z.phys.Chem., 153, 376 (1931).
37. F.R.DUKE and R.C.PINKERTON, J.Am.Chem.Soc., 73, 3045 (1951).

Chlorate Complexes

Complex ion	Temp. °C	Ionic strength	Method	k	pk	K	pK	Reference
$BaClO_3^+$	25	0	El.cond.	0.20	0.70	0.20	0.70	[1]
$ThClO_3^{3+}$	25	0.5	Distrib.	0.55	0.26	0.55	0.26	[2]

REFERENCES

1. T. MacDOUGALL and C. W. DAVIES, J. Chem. Soc., 1416 (1935).
2. R. A. DAY and R. W. STOUGHTON, J. Am. Chem. Soc., 72, 5662 (1950).

Iodate Complexes

Complex ion	Temp. °C	Ionic strength	Method	k	pk	K	pK	Reference
$BaJO_3^+$	25	0	El.cond. Solub.	$8.9 \cdot 10^{-2}$	1.05	$8.9 \cdot 10^{-2}$	1.05	[1]
$CaJO_3^+$	25	0	Do.	0.129	0.89	0.129	0.89	[2]
KJO_3	25	0	El.cond.	2.0	−0.30	2.0	−0.30	[3]
$MgJO_3^+$	25	0	Solub.	0.190	0.72	0.190	0.72	[4]
$SrJO_3^+$	25	0	El.cond. Solub.	0.10	1.00	0.10	1.00	[5]
$ThJO_3^{3+}$	25	0.5	Distrib.	$1.32 \cdot 10^{-3}$	2.88	$1.32 \cdot 10^{-3}$	2.88	[6]
$Th(JO_3)_2^{2+}$	25	0.5	"	$1.23 \cdot 10^{-2}$	1.91	$1.62 \cdot 10^{-5}$	4.79	[6]
$Th(JO_3)_3^+$	25	0.5	"	$4.4 \cdot 10^{-3}$	2.36	$7.1 \cdot 10^{-8}$	7.15	[6]
$TlJO_3$	25	0	"	0.32	0.50	0.32	0.50	[4]

REFERENCES

1. T. MacDOUGALL and C. W. DAVIES, J. Chem. Soc., 1416 (1935).
2. C. W. DAVIES, ibid., 271 (1938).
3. C. W. DAVIES, Trans. Farad. Soc., 26, 592 (1930).
4. C. W. DAVIES, J. Chem. Soc., 2410 (1930).
5. C. A. COLMAN-PORTER and C. B. MONK, ibid., 1312 (1952).
6. R. A. DAY and R. W. STOUGHTON, J. Am. Chem. Soc., 72, 5662 (1950).

Chloride Complexes

Complex ion	Temp. °C	Ionic strength	Method	k	pk	K	pK	References principal	supplementary
Ag_2Cl^+	room	—	Solub.	—	—	$2 \cdot 10^{-7}$	6.70	[1]	[33]
$AgCl$	25	0.0	"	$2.04 \cdot 10^{-9}$	2.69	$2.04 \cdot 10^{-3}$	2.69	[2]	[3,33]
$AgCl_2^-$	25	0.0	"	$8.7 \cdot 10^{-8}$	2.06	$1.76 \cdot 10^{-5}$	4.75	[2]	[34,35,36]
$AgCl_3^{2-}$	25	5.0	Potent.	—	—	$4.0 \cdot 10^{-6}$	5.40	[3]	[37]
$AgCl_4^{3-}$	25	0.0	Solub.	—	—	$1.2 \cdot 10^{-6}$	5.92	[4]	[38]
"	25	5.0	Potent.	1.2	−0.08	$4.8 \cdot 10^{-6}$	5.32	[3]	[37]
$AuCl_4^-$	18	—	—	—	—	$5 \cdot 10^{-22}$	21.30	[5]	—
$BiCl^{2+}$	25	1.0 — 2.0	Solub.	$3.6 \cdot 10^{-3}$	2.44	$3.6 \cdot 10^{-3}$	2.44	[6]	[39]
$BiCl_2^+$	25	1.0 — 2.0	"	0.22	0.66	$7.9 \cdot 10^{-4}$	3.10	[6]	[39]
$BiCl_3$	25	1.0 — 2.0	"	0.23	0.64	$1.8 \cdot 10^{-4}$	3.74	[6]	[39]
$BiCl_4^-$	25	1.0 — 2.0	"	0.94	0.03	$1.7 \cdot 10^{-4}$	3.77	[6]	[7,39]
$BiCl_5^{2-}$	18	2.5	Potent.	—	—	$(2 \cdot 10^{-6})$	(5.7)	[7]	—
$BiCl_6^{3-}$	18	1.6	"	—	—	$(3.8 \cdot 10^{-7})$	(6.4)	[7]	—
$CdCl^+$	25	1.0	"	$4.5 \cdot 10^{-2}$	1.35	$4.5 \cdot 10^{-2}$	1.35	[8]	[11,40,41]
$CdCl_2$	25	3.0	Polarog.	$2.86 \cdot 10^{-2}$	1.54	$2.86 \cdot 10^{-2}$	1.54	[9]	[42]
"	25	1.0	Potent.	$3.7 \cdot 10^{-1}$	0.43	$1.67 \cdot 10^{-2}$	1.78	[8]	[11,40,41]
$CdCl_3^-$	25	3.0	Polarog.	$3.0 \cdot 10^{-1}$	0.52	$8.71 \cdot 10^{-3}$	2.06	[9]	[42]
"	25	3.0	Polarog.	0.41	0.40	$3.40 \cdot 10^{-3}$	2.46	[9]	[40,41,42,43]
$CdCl_4^{2-}$	18	1.0—1.6	Solub.	—	—	$(9.3 \cdot 10^{-3})$	(2.0)	[10]	[41,42,43]
$CdCl_6^{4-}$	25	0.0	Polarog.	—	—	$(3.8 \cdot 10^{-3})$	(2.58)	[11]	[43]

Chloride complexes (contd.)

Complex ion	Temp. °C	Ionic strength	Method	k	pk	$\varkappa$	p$\varkappa$	References Principal	References supplementary
CeCl²⁺	18	0.0	Ion exch.	0.33	0.48	0.33	0.48	[12]	—
CrCl₂⁺	room	0.0	El.cond.	—	—	$1.26\cdot10^{-2}$	1.90	[13]	[44]
CuCl₂⁻	18	0.67	Potent.	—	—	$5.01\cdot10^{-6}$	5 30	[14]	[23,45]
CuCl⁺	25.2	1.0	Spectr.	0.77	0.11	0.77	−0 52	[15]	[46—51]
CuCl₂	25.2	1.0	.	4.3	−0.63	3.3	−0 52	[15]	[46—48,51]
FeCl²⁺	25	0.0	.	$3.3\cdot10^{-2}$	1.48	$3.3\cdot10^{-2}$	1.48	[16]	[52]
.	26.7	1.0	.	0.24	0.62	0.24	0.62	[16]	[53,54]
.	20	2.0	.	0.175	0.76	0.175	0.76	[17]	—
FeCl₂⁺	25	0.0	.	0.22	0.65	$7.4\cdot10^{-3}$	2.13	[6]	—
.	26.7	1.0	.	0.76	0.12	$1.8\cdot10^{-1}$	0 74	[16]	—
.	20	2.0	.	0.50	0.30	$8.7\cdot10^{-2}$	1.06	[17]	[17]
FeCl₃	25	0.0	.	10	−1.00	$7.4\cdot10^{-2}$	1.13	[16]	[17]
.	26.7	1.0	.	25	−1.40	4.6	−0.66	[16]	—
GaCl²⁺	—	0.5	—	1	0	1	0	[18]	—
HgCl⁺	25	0.5	Potent.	$5.4\cdot10^{-6}$	5.27	$5.4\cdot10^{-6}$	5.27	[19]	[55]
H Cl₂	25	0.5	.	$3.1\cdot10^{-8}$	7.51	$1.7\cdot10^{-18}$	12.78	[19]	[55,56]
HgCl₃⁻	25	0.5	.	$7.2\cdot10^{-2}$	1.14	$1.2\cdot10^{-14}$	13.92	[19]	[55,57]
HgCl₄²⁻	25	0.5	.	0.10	1.0	$1.2\cdot10^{-15}$	14.92	[19]	[56]
InCl²⁺	25	1.0	Ion exch.	$3.8\cdot10^{-2}$	1.42	$3.8\cdot10^{-2}$	1.42	[20]	[58]
InCl₂⁺	25	1.0	"	0.15	0.81	$5.9\cdot10^{-3}$	2.23	[20]	[58]
InCl₃	25	1.0	.	0.10	1.00	$5.9\cdot10^{-4}$	3.23	[20]	—
MnCl²⁺	25.2	2.0	Kin.	0.11	0.96	0.11	0.96	[21]	—
PbCl⁺	25	0.0	Polarog.	$2.3\cdot10^{-2}$	1.64	$2.3\cdot10^{-2}$	1.64	[11]	[59,60,67]

Chloride complexes (contd.)

Complex ion	Temp. °C	Ionic strength	Method	k	pk	K	pK	References principal	References supplementary
$PbCl_3^-$	25	0.0	Polarog.	—	—	$1.4\cdot10^{-2}$	1.85	[11]	[43]
$PdCl_4^{2-}$	25	1.0—4.0	Potent.	—	—	$6\cdot10^{-14}$	13,22	[22]	[23]
$PtCl_4^{2-}$	—	—		—		10^{-16}	16	[23]	—
$PuCl^{3+}$	25	2.00	Spectr.	2.5	-0.40	2.5	-0.40	[24]	—
$SnCl^+$	25	3.00	Potent.	$8.9\cdot10^{-2}$	1.05	$8.9\cdot10^{-2}$	1.05	[25]	[61,62]
$SnCl_2$	25	3.0	.	0.22	0.65	$2.0\cdot10^{-2}$	1.70	[25]	[62]
$SnCl_3^-$	25	3.0	.	1.05	0.02	$2.1\cdot10^{-2}$	1 68	[25]	[62]
$ThCl^{3+}$	25	0.5	Distrib.	0.57	0.25	0.57	0.25	[26]	—
$ThCl^{3+}$	25	4.0	Distrib.	0.77	0.12	0.77	0.12	[27]	[63]
$\cdot ThCl_2^{2+}$	25	4.0	Distrib.	10.5	-1.02	8.00	-0.90	[27]	[63]
$ThCl_3^+$	25	4.0	.	3.4	-0.53	27	-1.43	[27]	[63]
$ThCl_4$	25	4.0	.	2.6	-0.42	71	-1.85	[27]	[63]
$TlCl$	25	0.0	Solub.	0.21	0.68	0.21	0.68	[28]	[67]
$TlCl^{2+}$	25	0.1—0.2	Potent.	$8.0\cdot10^{-9}$	8.1	$8.0\cdot10^{-9}$	8.10	[29]	—
$TlCl_2^+$	25	0.1—0.2	.	$3.2\cdot10^{-6}$	5.5	$2.5\cdot10^{-21}$	13.60	[29]	—
$TlCl_3$	25	0.1—0.2	.	$6.3\cdot10^{-8}$	2.2	$1.6\cdot10^{-16}$	15.80	[29]	—
$TlCl_4^-$	25	0.1—0.2	.	$2.5\cdot10^{-8}$	2.6	$4\cdot10^{-10}$	18 40	[29]	—
UCl^{3+}	25	0.0	Spectr.	0.143	0.85	0.143	0.85	[30]	[65]
UO_2Cl^+	20	0.5	.	1.6	-0.20	1.6	-0.20	[30]	[67]
$ZnCl^+$	25	—	.	2.0	-0.30	2.0	-0.30	[31]	—
$ZnCl_2$	25	3.0	Potent.	1.54	-0.19	1.54	-0.19	[32]	[68]
$ZnCl_3^-$	25	3.0	.	2.6	-0.41	4.00	-0.60	[32]	—
$ZnCl_4^{2-}$	25	3.0	.	0.18	0.75	0.71	0.15	[32]	—

REFERENCES

1. K.B.YATSIMIRSKII, Dokl.Akad.Nauk SSSR, 77, 819 (1951);
    K.HELLWIG, Z.anorg.Chem., 25, 157 (1900).
2. J.H.JONTE and D.S.MARTIN, J.Am.Chem.Soc., 74, 2052 (1952).
3. E.BERNE and I.LEDEN, Svensk.Kem.Tidsk., 65, 88 (1953);
    Chem.Abs., 47, 10392 (1953).
4. W.ERBER and A.SCHULY, Chem.Abs., 35, 7801 (1941); J.
    Prakt.Chem., 158, 176 (1941).
5. N.BJERRUM and A.KIRSCHNER, Kgl.Danske Videnskab., V, No.
    1 (1918).
6. K.B.YATSIMIRSKII, Collected Articles on General Chemistry
    (Sbornik statei po obshchei khimii), I, 97, Izd.Akad.
    Nauk SSSR (1953).
7. A.K.BABKO and A.M.GOLUB, ibid., I, 64.
8. C.E.VANDERZEE and H.J.DAWSON, J.Am.Chem.Soc., 75, 5659
    (1953).
9. L.ERIKSSON, Acta Chem.Scand., 7, 1146 (1953); Chem.Abs.,
    48, 13366 (1954).
10. I.M.KORENMAN, Zh.obshch.khim., 18, 1233 (1948).
11. A.M.VASIL'EV and V.I.PROUKHINA, Zh.anal.khim., 6, 218
    (1951).
12. S.W.MAYER and S.D.SCHWARTZ, J.Am.Chem.Soc., 73, 222 (1951).
13. A.B.LAMB and G.R.FONDA, ibid., 43, 1155 (1921).
14. A.I.STABROVSKII, Zh.fiz.khim., 26, 949 (1952).
15. H.McCONNELL and N.DAVIDSON, J.Am.Chem.Soc., 72, 3164 (1950).
16. E.RABINOWITCH and W.STOCKMAYER, ibid., 64, 335 (1942).
17. G.A.GAMLEN and D.O.JORDAN, J.Chem.Soc., 1435 (1953).
18. H.TAUBE and A.S.WILSON, cited in W.M.Latimer,Okislitel'-
    nye sostoyaniya elementov i ikh potentsialy v vodnykh
    restvorakh (The Oxidation States of the Elements and
    their Potentials in Aqueous Solution) (Translation
    into Russian), IL (1954).
19. A.JOHNSON, J.QUARFORT and L.G.SILLEN, Chem.Abs., 42, 2161
    (1948).
20. J.A.SCHUFFE and H.M.EILAND, J.Am.Chem.Soc., 76, 960 (1954).
21. H.TAUBE, ibid., 70, 3928 (1948).
22. D.H.TEMPLETON, G.W.WATT and C.S.GARNER, ibid., 65, 1608
    (1943).

23. W.M.LATIMER, Okislitel'nye sostoyaniya elementov i ikh
    potentsialy v vodnykh restvorakh (The Oxidation States
    of the Elements and Their Potentials in Aqueous Solu-
    tion) (Translation into Russian), IL (1954).
24. J.C.HINDMAN, Chem.Abs., 44, 3831 (1950).
25. C.E.VANDERZEE and D.E.RHODES, J.Am.Chem.Soc., 74, 3552
    (1952).
26. R.A.DAY and R.W.STOUGHTON, ibid., 72, 5662 (1950).
27. E.L.ZEBROSKI, H.W.ALTER and F.K.NEUMANN, ibid., 73, 5646
    (1951).
28. R.P.BELL and J.H.B.GEORGE, Trans.Farad.Soc., 49, 619
    (1953).
29. R.BENOIT, Bull.Soc.Chim.France, 518 (1949); Chem.Abs.,
    43, 8939 (1949).
30. K.A.KRAUS and F.NELSON, J.Am.Chem.Soc., 72, 3901 (1950).
31. S.AHRLAND, Chem.Abs., 47, 1528 (1953).
32. L.G.SILLEN and B.LILJEQVIST, ibid., 40, 4588 (1946).
33. A.PINKUS, S.FREDERIK and R.SCHEPMANS, Bull.Soc.Chim.
    Belges, 47, 304 (1937); Zbl., II, 1382 (1938).
34. I.M.KORENMAN, Zh.obshch.khim., 16, 157 (1946).
35. J.E.BARNEY, W.J.ARGERSINGER and C.A.REYNOLDS, J.Am.Chem.
    Soc., 73, 3785 (1951).
36. G.S.FORBES, ibid., 33, 1937 (1911).
37. I.LEDEN, Svensk.Kem.Tidsk., 64, 249 (1952); Chem.Abs.,
    48, 3114 (1954).
38. G.BODLANDER and W.EBERLEIN, Z.anorg.Chem., 39, 197 (1904)
39. A.K.BABKO, Naukovi Zapiski KDU, 4, 81 (1939).
40. E.L.KING, J.Am.Chem.Soc., 71, 319 (1949).
41. H.L.RILEY and V.GALLAFENT, J.Chem.Soc., 514 (1932).
42. I.LEDEN, Z.Phys.Chem., A, 188, 160 (1941).
43. I.A.KORSHUNOV, N.I.MALYUGINA and O.M.BALABANOVA, Zh.
    obshch.khim., 21, 620 (1951).
44. S.G.SHUTTLEWORTH, J.Soc.Leather Trades Chemists, 38, 110
    (1954); Ref. zh. khim., 23513 (1955).
45. A.A.NOYES and M.J.SHOW, J.Am.Chem.Soc., 40, 739 (1948).
46. J.BJERRUM, Kgl.Danske Vidensk.Selskab., No.18, 22 (1946).
47. A.K.BABKO, Naukovi Zapiski KDU, 4, 81 (1939).
48. H.L.RILEY and H.C.SMITH, J.Chem.Soc., 1448 (1934).
49. C.B.MONK, Trans.Farad.Soc., 47, 285 (1951).
50. R.NASANEN, Suomen.Kemist, B 26, 37 (1953); Ref.zh.khim.,
    11487 (1955).
51. R.KRUH, J.Am.Chem.Soc., 76, 4865 (1954).
52. W.C.BRAY and A.V.KERSHEY, ibid., 56, 1889 (1931).
53. C.BROSSET, Svensk.Kem.Tidsk., 53, 434 (1941).

54. J.BADOZ-LAMBLING, Bull.Soc.Chim.France., 552 (1950).
55. H.MORZE, Z.Phys.Chem., 41, 709 (1902).
56. M.S.SCHERILL, ibid., 43, 705 (1903); 47, 103 (1904).
57. D.PESCHANSKI, J.Chim.Phys., 50, 640 (1953); Chem.Abs., 48, 4939 (1954).
58. M.SUNDEN, Svensk.Kem.Tidsk., 66, 20 (1954); 66, 173 (1954); Chem.Abs., 48, 9252 (1954); 43, 1465 (1955).
59. M.CAVIGLI, Chem.Abs., 46, 3832 (1952).
60. H. FROMHERZ, Z.phys.Chem., 153, 382 (1931).
61. F.R.DUKE and R.C.PINKERTON, J.Am.Chem.Soc., 73, 3045 (1951).
62. F.R.DUKE and W.G.COURTENAY, cited ibid., 74, 3552 (1952).
63. W.C.WAGGENER and R.W.STOUGHTON, J.Phys.Chem., 56, 1 (1952).
64. KUO-HAO HU and A.B.SCOTT, J.Am.Chem.Soc., 77, 1380 (1955).
65. S.AHRLAND and R.LARSSON, Acta Chem.Scand., 8, 137 (1954); Chem.Abs., 48, 11969 (1954).
66. R.A.DAY, R.N.WILHITE and F.D.HAMILTON, J.Am.Chem.Soc., 77, 3180 (1955).
67. G.H.NANCOLLAS, J.Chem.Soc., 1458 (1955).
68. E.FERRELL, ibid., 1124 (1936).

Hydroperoxo-Complexes

Complex ion	Temp. °C	Ionic Strength	Method	k	pk	K	pK	References principal	References supplementary
$Fe(O_2H)^{2+}$	20	0.1	Spectr.	$5.0 \cdot 10^{-10}$	9.30	$5.0 \cdot 10^{-10}$	9.30	[1]	—
$TiO(H_2O_2)^{2+}$	Room	0.1	"	$1.0 \cdot 10^{-4}$	4.00	$1.0 \cdot 10^{-4}$	4.00	[2]	[3,4]

REFERENCES

1. M.G.EVANS, P.GEORGE and N.URI, Trans.Farad.Soc., 45, 230 (1949).
2. K.E.KLEINER, Zh.obshch.khim., 22, 17 (1952).
3. A.K.BABKO and A.I.VOKOVA, ibid., 21, 1949 (1951).
4. E.GASTINGER, Z.anorg.Chem., 275, 331 (1954).

Cyanide Complexes

Complex ion	Temp. °C	Ionic strength	Method	k	pk	K	pK	References principal	References supplementary
Ag(CN)$_2^-$	18	0.3	Potent.	—	—	$8 \cdot 10^{-22}$	21.1	[1]	[4,10,11]
Ag(CN)$_3^{2-}$	25	0.1—1	Spectr.	0.113	0.95	—	—	[2]	[10]
.	25	0	.	0.20	0.70	$1.6 \cdot 10^{-22}$	21.8	[2]	—
Ag(CN)$_4^{3-}$	25	0.1—1	.	3 18	−0.50	—	—	[2]	—
.	25	0	.	13.4	−1.13	$2\,1 \cdot 10^{-21}$	20.68	[2]	—
Au(CN)$_2^-$	—	—	Potent.	—	—	$5.0 \cdot 10^{-39}$	38.3	[3,4]	—
CdCN$^+$	25	3.0	.	$2.9 \cdot 10^{-6}$	5.54	$2.9 \cdot 10^{-6}$	5.54	[5]	—
Cd(CN)$_2$	25	3.0	.	$8.7 \cdot 10^{-6}$	5.06	$2.5 \cdot 10^{-11}$	10.60	[5]	—
Cd(CN)$_3^-$	25	3.0	.	$2.0 \cdot 10^{-5}$	4.70	$5.0 \cdot 10^{-16}$	15.30	[5]	[9,12]
Cd(CN)$_4^{2-}$	25	3.0	.	$2\,8 \cdot 10^{-4}$	3.55	$1.41 \cdot 10^{-19}$	18.85	[5]	[9,10,13,14]
Cu(CN)$_2^-$	25	0	.	—	—	$1 \cdot 10^{-24}$	24.0	[6]	[4]
Cu(CN)$_3^{2-}$	25	0	Spectr.	$2.6 \cdot 10^{-5}$	4.59	$2.6 \cdot 10^{-29}$	28.59	[7]	[15]
Cu(CN)$_4^{3-}$	25	0	.	$2 \cdot 10^{-1}$	1.70	$5.0 \cdot 10^{-33}$	30.30	[7]	[13]
Fe(CN)$_6^{4-}$	18	—	Potent.	—	—	10^{-36}	35	[4]	—
Fe(CN)$_6^{3-}$	18	—	.	—	—	10^{-42}	42	[4]	—
Hg(CN)$_4^{2-}$	25	0.05—0.20	.	—	—	$4.0 \cdot 10^{-42}$	41.4	[8]	—
Zn(CN)$_4^{2-}$	18	0.1—0.2	.	—	—	$1.3 \cdot 10^{-17}$	16.89	[9]	[12,13,15,16]

REFERENCES

1. R. GAUGIN, J. Chim. Phys., 42, 28 (1945).
2. L. H. JONES and R. A. PENNEMAN, J. Chem. Phys., 22, 965 (1954).
3. G. BODLANDER, Ber., 36, 3933 (1903).
4. W. M. LATIMER, Okislitel'nye sostoyaniya elementov i ikh potentsialy v vodnykh rastvorakh (The Oxidation States of the Elements and their Potentials in Aqueous Solution) (Translation into Russian), IL (1954).
5. I. LEDEN, Svensk. Kem. Tidsk., 56, 31 (1944); Chem. Abs., 40, 3070 (1946).
6. M. G. VLADIMIROV and I. A. KAKOVSKII, Zh. prikl. khim., 23, 580 (1950).
7. R. A. PENNEMAN and L. H. JONES, J. Chem. Phys., 24, 293 (1956).
8. M. S. SHERILL, Z. Phys. Chem., 43, 705 (1903); 47, 103 (1904).
9. H. EULER, Ber., 36, 3400 (1903).
10. E. FERRELL et al., J. Chem. Soc., 1121 (1936).
11. R. GAUGIN, Ann. chim., 4, 832 (1949); Chem. Abs., 44, 4757 (1950).
12. J. HEYROVSKY, Polyarograficheskii metod (Polarography) (Translation into Russian), Ob'ed. nauchno-tekh. izd. (1937).
13. S. SUZUKI, Chem. Abs., 48, 7491 (1954).
14. F. HIRATA, ibid., 46, 1382 (1952).
15. A. I. STABROVSKII, Zh. fiz. khim., 26, 949 (1952).
16. F. KUNSCHERT, Z. anorg. Chem., 41, 341 (1904).

Fluoride Complexes

Complex ion	Temp. °C	Ionic strength	Method	k	pk	K	pK	References principal	References supplementary
AlF²⁺	25	0.53	Potent.	$7.4 \cdot 10^{-7}$	6.13	$7.4 \cdot 10^{-7}$	6.13	[1]	[15]
AlF₂⁺	25	0.53	.	$9.5 \cdot 10^{-6}$	5.02	$7.1 \cdot 10^{-12}$	11.15	[1]	—
AlF₃	25	0.53	.	$1.4 \cdot 10^{-4}$	3.85	$1.0 \cdot 10^{-15}$	15.00	[1]	—
AlF₄⁻	25	0.53	.	$1.8 \cdot 10^{-3}$	2.75	$1.8 \cdot 10^{-18}$	17.75	[1]	—
AlF₅²⁻	25	0.53	.	$2.4 \cdot 10^{-2}$	1.62	$4.3 \cdot 10^{-20}$	19.37	[1]	—
AlF₆³⁻	25	0.53	.	$3.4 \cdot 10^{-1}$	0.47	$1.44 \cdot 10^{-20}$	19.84	[1]	—
BeF⁺	25	0.01—6	Solub.	$5.1 \cdot 10^{-6}$	4.29	$5.1 \cdot 10^{-5}$	4.29	[2]	[10,16]
BeF₂	25	0.01—6	.	$1.0 \cdot 10^{-2}$	2.00	$5.1 \cdot 10^{-7}$	6.29	[2]	—
CeF²⁺	25	0	Potent.	$6.3 \cdot 10^{-4}$	3.20	$6.3 \cdot 10^{-4}$	3.20	[3]	[17]
CrF²⁺	25	0.5	Spectr.	$3.9 \cdot 10^{-5}$	4.41	$3.9 \cdot 10^{-5}$	4.41	[4]	—
CrF₂⁺	25	0.5	.	$4.0 \cdot 10^{-4}$	3.40	$1.5 \cdot 10^{-8}$	7.81	[4]	—
CrF₃	25	0.5	.	$3.3 \cdot 10^{-3}$	2.48	$5.1 \cdot 10^{-11}$	10.29	[4]	—
FeF²⁺	25	0.5	Potent.	$5.2 \cdot 10^{-6}$	5.28	$5.2 \cdot 10^{-6}$	5.28	[5]	[17,18,19]
FeF₂⁺	25	0.5	.	$9.5 \cdot 10^{-5}$	4.02	$5.0 \cdot 10^{-10}$	9.30	[5]	[18]
FeF₃	25	0.5	.	$1.7 \cdot 10^{-3}$	2.76	$8.7 \cdot 10^{-13}$	12.06	[5]	—
GaF²⁺	25	0.5	Spectr.	$8.34 \cdot 10^{-6}$	5.08	$8.34 \cdot 10^{-6}$	5.08	[4]	—
GdF²⁺	25	0.5	Potent.	$3.5 \cdot 10^{-4}$	3.46	$3.5 \cdot 10^{-4}$	3.46	[3]	—
HF	25	0	.	$1.0 \cdot 10^{-3}$	3.0	$1.0 \cdot 10^{-3}$	3.0	[10]	—
InF²⁺	20	1.0	.	$2.0 \cdot 10^{-4}$	3.70	$2.0 \cdot 10^{-4}$	3.70	[6]	[20,21]
InF₂⁺	20	1.0	.	$2.8 \cdot 10^{-3}$	2.55	$5.6 \cdot 10^{-7}$	6.25	[6]	[21,22]
InF₃	20	1.0	.	$4.5 \cdot 10^{-3}$	2.35	$2.5 \cdot 10^{-9}$	8.60	[6]	[21]
InF₄⁻	20	1.0	.	$7.9 \cdot 10^{-2}$	1.10	$2.0 \cdot 10^{-10}$	9.70	[6]	—

Fluoride complexes (contd.)

Complex ion	Temp. °C	Ionic strength	Method	k	pk	K	p^K	References prin-cipal	References supple-mentary
LaF²⁺	25	0.0	Potent.	$1.7\cdot10^{-3}$	2.77	$1.7\cdot10^{-3}$	2.77	[3]	—
MgF⁺	25	0.5	Kin.	$5.0\cdot10^{-2}$	1.30	$5.0\cdot10^{-2}$	1.30	[7]	—
MnF²⁺	25.2	2.0		$3.3\cdot10^{-6}$	5.48	$3.3\cdot10^{-6}$	5.48	[8]	—
PuF³⁺	25	2.0	Potent.	$1.7\cdot10^{-7}$	6.77	$1.7\cdot10^{-7}$	6.77	[9]	—
ScF₆³⁻	—	—	—	—	—	$5\cdot10^{-18}$	17.3	[10]	—
ThF³⁺	25	0.5	Potent.	$2.2\cdot10^{-8}$	7.65	$2.2\cdot10^{-8}$	7.65	[5]	[22,23]
ThF₂²⁺	25	0.5		$1.5\cdot10^{-6}$	5.81	$3.5\cdot10^{-14}$	13.46	[5]	[22,23]
ThF₃⁺	25	0.5	Spectr.	$3.1\cdot10^{-5}$	4.51	$1.1\cdot10^{-16}$	17.97	[5]	—
TiOF	18	0.1	Solub.	$3.6\cdot10^{-7}$	6.44	$3.6\cdot10^{-7}$	6.44	[11]	—
TlF	25	0.0	Potent.	0.8	0.1	0.8	0.1	[12]	—
UO₂F⁺	20	—		$2.6\cdot10^{-5}$	4.59	$2.6\cdot10^{-5}$	4.59	[13]	—
UO₂F₂	20	—		$4.6\cdot10^{-4}$	3.34	$1.2\cdot10^{-8}$	7.93	[13]	—
UO₂F₃⁻	20	—		$2.9\cdot10^{-3}$	2.54	$3.4\cdot10^{-11}$	10.47	[13]	—
UO₂F₄²⁻	20	—	Distrib.	$4.3\cdot10^{-2}$	1.37	$1.4\cdot10^{-2}$	11.84	[13]	—
ZrF³⁺	25	2.0		$1.6\cdot10^{-9}$	8.80	$1.6\cdot10^{-9}$	8.80	[14]	—
ZrF₂²⁺	25	2.0		$4.8\cdot10^{-8}$	7.32	$7.6\cdot10^{-17}$	16.12	[14]	—
ZrF₃⁺	25	2.0		$1.5\cdot10^{-6}$	5.82	$1.2\cdot10^{-22}$	21.94	[14]	—

REFERENCES

1. C. BROSSET and J. ORRING, Svensk. Kem. Tidsk., 55, 101 (1943); Chem. Abs., 37, 24 (1945).
2. I. V. TANANAYEV and E. N. DEICHMAN, Izv. Akad. Nauk SSSR, Otd. khim. nauk, 144 (1949).
3. J. W. KURY, Chem. Abs., 48, 3833 (1954).
4. A. S. WILSON and H. TAUBE, J. Am. Chem., 74, 3509 (1952).
5. H. W. DODGEN and G. K. ROLLEFSON, ibid., 71, 2600 (1949).
6. N. SUNDEN, Svensk. Kem. Tidsk., 66, 50 (1954); Ref. zh. khim., 18547 (1955).
7. R. E. CONNICK, Maak-Sang TSAO, J. Am. Chem. Soc., 76, 5311 (1954).
8. H. TAUBE, ibid., 70, 3928 (1948).
9. C. K. McLANE, Chem. Abs., 44, 3831 (1950).
10. W. M. LATIMER, Okislitel'nye sostoyaniya elementov i ikh potentsialy v vodnykh rastvorakh (The Oxidation States of the Elements and their Potentials in Aqueous Solution) (Translation into Russian), IL (1954).
11. K. E. KLEINER, Zh. obshch. khim., 22, 17 (1952).
12. R. P. BELL and J. H. B. GEORGE, Trans. Farad. Soc., 49, 619 (1953).
13. S. AHRLAND and R. LARSSON, Acta Chem. Scand., 8, 354 (1954); Chem. Abs., 48, 11970 (1954).
14. R. E. CONNICK and W. H. McVEY, J. Am. Chem. Soc., 71, 3182 (1949).
15. K. E. KLEINER, Zh. obshch. khim., 20, 1747 (1950).
16. K. E. KLEINER, ibid., 21, 18 (1951).
17. S. W. MAYER and S. D. SCHWARTZ, J. Am. Chem. Soc., 73, 390 (1951).
18. A. K. BABKO and E. K. KLEINER, Zh. obshch. khim., 17, 1259 (1947).
19. C. BROSSET, Chem. Abs., 37, 24 (1943).
20. G. SAINI, Gazz. Chim. Ital., 83, 677 (1953); Chem. Abs., 48, 6899 (1954).
21. L. G. HEPLER, J. W. KURY and Z. Z. HUGUS, J. Phys. Chem., 58, 26 (1954).
22. J. A. SCHUFLE and H. M. EILAND, J. Am. Chem. Soc., 76, 960 (1954).
23. R. A. DAY and R. W. STOUGHTON, ibid., 72, 5662 (1950).
24. E. L. ZEBROSKI, H. W. ALTER and F. K. NEUMAN, ibid., 73, 5646 (1951).

Hydroxy-Complexes

Complex ion	Temp. °C	Ionic strength	Method	k	pk	K	pK	References principal	References supplementary
$AgOH$	25	—	Potent.	$5.0 \cdot 10^{-3}$	2.30	$5.0 \cdot 10^{-3}$	2.30	[1]	—
$AlOH^{3+}$	25	0	Potent.,El.cond.	$1.38 \cdot 10^{-9}$	8.86	$1.38 \cdot 10^{-9}$	8.86	[2]	[8]
$BaOH^{+}$	25	0	Kin.	0.23	0.64	0.23	0.64	[3]	[20]
$BeOH^{+}$	25	1.0	Potent.	$3.3 \cdot 10^{-8}$	7.50	$3.3 \cdot 10^{-8}$	7.50	[4]	—
Be_2OH^{3+}	25	1.0	Kin.			$3.3 \cdot 10^{-11}$	10.50	[4]	—
$CaOH^{+}$	30	0	Potent.	$5.0 \cdot 10^{-2}$	1.30	$5.0 \cdot 10^{-2}$	1.30	[3]	[20,28]
$CdOH^{+}$	25	0.1	Potent.	$5.0 \cdot 10^{-3}$	2.30	$5.0 \cdot 10^{-3}$	2.30	[5]	—
$CeOH^{3+}$	25	Variable	Spectr.	$2.0 \cdot 10^{-15}$	14.70	$2.0 \cdot 10^{-15}$	14.70	[6]	—
$CoOH^{+}$	25	0	Potent., Calc.	$4 \cdot 10^{-5}$	4.4	$4 \cdot 10^{-5}$	4.4	[7]	[5,33]
$CrOH^{3+}$	25	0.005	El.cond,Potent.	$1.02 \cdot 10^{-10}$	9.99	$1.02 \cdot 10^{-10}$	9.99	[8]	[34]
$CuOH^{+}$	25	0	Potent.	$3.4 \cdot 10^{-7}$	6.47	$3.4 \cdot 10^{-7}$	6.47	[9]	[5,35]
$Cu(OH)_4^{2-}$	room	Variable	Solub.			$7 \cdot 6 \cdot 10^{-17}$	16.12	[10]	—
$FéOH^{+}$	25	0	Potent., Calc.	$1.3 \cdot 10^{-1}$	3.9	$1.3 \cdot 10^{-4}$	3.9	[7]	[36,37]
$FeOH^{2+}$	25	0	Potent., Calc.	$1.55 \cdot 10^{-12}$	11.81	$1.55 \cdot 10^{-12}$	11.81	[11]	[38]
"	25	0.1	Potent.	$7.9 \cdot 10^{-12}$	11.10	$7.9 \cdot 10^{-12}$	11.10	[12]	[39,40]
$Fe(OH)_2^{+}$	25	3.0	Potent.	$1.12 \cdot 10^{-11}$	10.95	$1.12 \cdot 10^{-11}$	10.95	[13]	[5]
$Fe_2(OH)_2^{4+}$	25	0.1	Spectr.	$1.82 \cdot 10^{-11}$	10.74	$2.04 \cdot 10^{-22}$	21.69	[13]	[40]
	25	0.0				$8.0 \cdot 10^{-26}$	25.10	[14]	[13,41]
	25	3 0				$8.0 \cdot 10^{-26}$	25.10	[14]	—
$GaOH^{3+}$	25	Variable		$2.5 \cdot 10^{-11}$	10 60	$2.5 \cdot 10^{-11}$	10.60	[15]	[42]
$HgOH^{+}$	25	0.5	Potent.	$5.0 \cdot 10^{-11}$	10.30	$5.0 \cdot 10^{-11}$	10.30	[16]	—
$Hg(OH)_2$	25	0.5	Potent.	$4.0 \cdot 10^{-12}$	11.40	$2.0 \cdot 10^{-22}$	21.70	[16]	[43]
$JnOH^{2+}$	23	0.006	Potent.	$5.0 \cdot 10^{-11}$	10.30	$5.0 \cdot 10^{-11}$	10.30	[17]	—
$Jn(OH)_4^{-}$	room	—	Polarog.			$2.5 \cdot 10^{-30}$	29.6	[18]	—

Hydroxy-complexes (contd.)

Complex ion	Temp. °C	Ionic Strength	Method	k	pk	K	pK	References principal	References supplementary
LaOH²⁺	25	0	Spectr.	$5.0\cdot10^{-4}$	3.30	$5.0\cdot10^{-4}$	3.30	[19]	—
LiOH	25	0	Potent.	0.66	0.18	0.66	0.18	[20]	[44]
MgOH⁺	25	0	.	$2.5\cdot10^{-3}$	2.58	$2.5\cdot10^{-3}$	2.58	[21]	[5]
MnOH⁺	30	0	.	$5.0\cdot10^{-4}$	3.30	$5.0\cdot10^{-4}$	3.30	[5]	—
NaOH	25	0	.	3	—0.48	3	—0.48	[20]	—
NiOH⁺	30	0.1	.	$2.5\cdot10^{-5}$	4.60	$2.5\cdot10^{-5}$	4.60	[5]	[33]
PbOH⁺	18	0	.	$6.0\cdot10^{-7}$	6.22	$6.0\cdot10^{-7}$	6.22	[22]	—
PrOH³⁺	25	1.0	.	$3.2\cdot10^{-13}$	12.49	$3.2\cdot10^{-13}$	12.49	[24]	[30]
PuOH³⁺	25	1.0	.	$7.6\cdot10^{-10}$	9.12	$7.6\cdot10^{-10}$	9.12	[25]	—
ScOH²⁺	25	3.0	.	$5.0\cdot10^{-13}$	12.30	$5.0\cdot10^{-13}$	12.30	[26]	[45,46]
SrOH⁺	25	0	.	0.15	0.82	0.15	0.82	[27]	[47]
SnOH⁺	25	1.0	.	$2.0\cdot10^{-10}$	9.70	$2.0\cdot10^{-10}$	9.70	[28]	[3]
ThOH³⁺	25	0	Potent.	0.15	0.82	0.15	0.82	[29]	[48,49]
TlOH	25	3.0	Solub.	$7.9\cdot10^{-15}$	14.10	$7.9\cdot10^{-15}$	14.10	[29]	—
TlOH²⁺	25	3.0	Potent.	$6.8\cdot10^{-15}$	14.17	$5.4\cdot10^{-29}$	28.27	[30]	—
Ti(OH)₂⁺	25	0.5	.	$4.8\cdot10^{-14}$	13.32	$4.8\cdot10^{-14}$	13.32	[30]	[50]
UOH³⁺	25	Variable	.	$3.2\cdot10^{-13}$	12.50	$3.2\cdot10^{-13}$	12.50	[31]	[51]
VOH²⁺	25	0	.	$8.3\cdot10^{-12}$	11.08	$8.3\cdot10^{-12}$	11.08	[31]	—
V(OH)₂⁺	25	Variable	.	$3.2\cdot10^{-11}$	10.50	$2.6\cdot10^{-21}$	21.58	[9]	[5]
ZnOH⁺	25	0	.	$4.0\cdot10^{-5}$	4.40	$4.0\cdot10^{-5}$	4.40	[18]	—
Zn(OH)₃⁻	room	—	Polarog.	—	—	$4.3\cdot10^{-15}$	14.37	—	—
Zn(OH)₄²⁻	25	Variable	Solub.	—	—	$3.6\cdot10^{-16}$	15.44	[32]	[52,53]

REFERENCES

1. J. BJERRUM, Chem. Rev., 46, 381 (1950).
2. T. ITO and N. YUI, Chem. Abs., 48, 5613 (1954).
3. R. P. BELL and J. E. PRUE, J. Chem. Soc., 362 (1949).
4. G. MATTOCK, J. Am. Chem. Soc., 76, 4835 (1954).
5. S. CHABEREOK et al., ibid., 74, 5057 (1952).
6. T. J. HARDWICK and E. ROBERTSON, Chem. Abs., 46, 3372 (1952).
7. K. B. YATSIMIRSKII, Zh. obshch. khim., 24, 1498 (1954).
8. N. BJERRUM, Z. phys. Chem., 59, 336 (1907).
9. B. B. OWEN and R. W. GURRY, J. Am. Chem. Soc., 60, 3074 (1938).
10. W. FREITKNECHT, Helv. chim. Acta, 27, 771 (1944).
11. T. H. SIDDAL and W. C. VOSBURGH, J. Am. Chem. Soc., 73, 4270 (1951).
12. T. V. ARDEN, J. Chem. Soc., 350 (1951).
13. B. HEDSTROM, Arkiv. Kemi., 6, 1 (1953); Chem. Abs., 47, 11939 (1953).
14. R. MILBURN and W. C. VOSBURGH, J. Am. Chem. Soc., 77, 1352 (1955).
15. T. MOELLER and G. L. KING, J. Phys. Colloid Chem., 54, 999 (1950).
16. S. HIETANEN and L. G. SILLEN, Chem. Abs., 47, 2577 (1953).
17. E. M. HATTOX and T. De VRIES, J. Am. Chem. Soc., 58, 2126 (1936).
18. J. HEYROVSKY, Polyarograficheskii metod (Polarography) (Translation into Russian), Ob'ed nauchno-tekh. izd., Leningrad (1937).
19. C. W. DAVIES, J. Chem. Soc., 1256 (1951); I. M. KOLTHOFF and R. ELMQUIST, J. Am. Chem. Soc., 53, 1217 (1931).
20. F. G. R. GIMBLET and C. B. MONK, Trans. Farad. Soc., 50, 965 (1954).
21. D. J. STOCK and C. W. DAVIES, ibid., 44, 856 (1948).
22. K. J. PEDERSEN, Chem. Abs., 40, 4588 (1946).
23. A. B. GARRETT, S. VELLENGA and C. M. FONTANA, J. Am. Chem. Soc., 61, 367 (1939).
24. S. W. RABIDEAU and J. F. LAMONS, ibid., 73, 2895 (1951).
25. M. KIRPATRICK and L. POKRAS, J. Electrochem. Soc., 100, 85 (1953); Chem. Abs., 47, 11963 (1953).
26. C. E. VANDERZEE and D. E. RHODES, J. Am. Chem. Soc., 74, 3552 (1952).
27. K. A. KRAUS and R. W. HOLMBERG, J. Phys. Chem., 58, 325 (1954).
28. R. P. BELL and J. H. B. GEORGE, Trans. Farad. Soc., 49, 619 (1953).

29. G. BIEDERMANN, Arkiv. Kemi, 5, 441 (1953); Chem. Abs., 47, 11938 (1953).
30. K. A. KRAUS and F. NELSON, J. Am. Chem. Soc., 72, 3901 (1950).
31. L. MEITES, ibid., 75, 6059 (1953); G. JONES and W. A. RAY, ibid., 66, 1571 (1944).
32. H. G. DIETRICH and J. JOHNSON, ibid., 49, 1419 (1927).
33. K. GAYER and L. WOONTNER, ibid., 74, 1436 (1952).
34. A. B. LAMB and G. R. FONDA, ibid., 43, 1155 (1921).
35. H. GUITER, Compt. rend., 228, 569 (1949).
36. D. L. LEUSSING and J. M. KOLTHOFF, J. Am. Chem. Soc., 75, 2476 (1953).
37. B. HEDSTROM, Arkiv Kemi, 5, 457 (1953); Chem. Abs., 47, 11938 (1953).
38. W. C. BRAY and A. V. HERSHEY, J. Am. Chem. Soc., 56, 1889 (1931).
39. C. BROSSET, Chem. Abs., 37, 24 (1943).
40. T. ITO and N. YUI, ibid., 48, 6791 (1954).
41. L. N. MULAY and P. W. SELWOOD, J. Am. Chem. Soc., 77, 2693 (1955).
42. R. FRICKE and K. MEYRING, Z. anorg. Chem., 176, 329 (1928)
43. A. B. GARRETT and A. E. HIRSCHLER, J. Am. Chem. Soc., 60, 299 (1938).
44. L. S. DARKEN and H. F. MEIER, ibid., 64, 621 (1942).
45. M. GORMAN, ibid., 61, 3342 (1939).
46. A. B. GARRETT and R. E. HEINS, ibid., 63, 562 (1941).
47. C. A. COLMAN-PORTER and C. B. MONK, J. Chem. Soc., 1312 (1952).
48. R. BENOIT, Bull. Soc. chim. France., 518 (1949); Chem. Abs., 8939 (1949).
49. C. E. JOHNSON, J. Am. Chem. Soc., 74, 959 (1952).
50. K. A. KRAUS and F. NELSON, ibid., 77, 3721 (1955).
51. S. C. FURMAN and C. S. GARNER, ibid., 72, 1785 (1950).
52. I. A. KORSHUNOV and E. F. KHRUL'KOVA, Zh. obshch. khim., 19, 2045 (1949).
53. M. STOCKELBERG and H. FREYHOLD, Z. Elektrochem., 46, 120 (1940).

Iodide Complexes

Complex ion	Temp °C	Ionic strength	Method	k	pk	K	pK	References principal	References supplementary
Ag_3J^{2+}	room	Variab.	Solub.	—	—	$8 \cdot 10^{-15}$	14.15	[1]	—
AgJ_3^{2-}	.	1.6	Potent.	—	—	$1.4 \cdot 10^{-14}$	13.95	[2]	[14,15]
AgJ_4^{3-}	.	1.6	.	—	—	$1.8 \cdot 10^{-14}$	13.75	[2]	[16,17]
BiJ_6^{3-}	20	Variab.	Potent.	—	—	$3.1 \cdot 10^{-12}$	11.54	[3]	—
Cd_2J^{3+}	25	0.2—9.0	Solub.	—	—	$3.2 \cdot 10^{-3}$	2.49	[4]	—
CdJ^+	25	0.0	Potent.	$5.2 \cdot 10^{-3}$	2.28	$5.2 \cdot 10^{-3}$	2.28	[5]	[4,18]
.	25	3.0	.	$8.3 \cdot 10^{-3}$	2.08	$8.3 \cdot 10^{-3}$	2.08	[6]	
CdJ_2	25	0.0	.	$2.3 \cdot 10^{-2}$	1.64	$1.2 \cdot 10^{-4}$	3.92	[5]	[18]
.	25	3.0	.	(0.24)	(0.62)	$(2 \cdot 0 \cdot 10^{-7})$	(2.70)	[6]	
CdJ_3^-	25	0.0	.	$(8.3 \cdot 10^{-2})$	(1.08)	$(1.0 \cdot 10^{-5})$	(5.0)	[5]	[7,18]
.	25	3.0	.	$5.0 \cdot 10^{-3}$	2.30	$1.0 \cdot 10^{-6}$	5.0	[6]	
CdJ_4^{2-}	25	0.0	.	$7.9 \cdot 10^{-2}$	1.10	$8 \cdot 10^{-7}$	6.10	[5]	[7,18, 19,20]
.	25	3.0	.	$3.2 \cdot 10^{-2}$	1.49	$3.0 \cdot 10^{-7}$	6.49	[6]	

Iodide complexes (contd.)

Complex ion	Temp. °C	Ionic strength	Method	k	pk	K	pK	References principal	References supplementary
CdJ_6^{4-}	25	0.05—2.5	Polarog.	—	—	$1.0 \cdot 10^{-6}$	6.0	[7]	—
CuJ_2^-	25	0.02—0.5	Potent.	—	—	$1.75 \cdot 10^{-9}$	8.76	[8]	—
Hg_2J^{3+}	25	0.05—3	Solub.	—	—	$1.77 \cdot 10^{-14}$	13.75	[9]	[9, 21]
HgJ^+	25	0.5	Potent.	$1.35 \cdot 10^{-13}$	12.87	$1.35 \cdot 10^{-13}$	12.87	[10]	[9, 21]
HgJ_2	25	0.5	"	$1.12 \cdot 10^{-11}$	10.95	$1.51 \cdot 10^{-24}$	23.82	[10]	[22]
HgJ_3^-	25	0.5	Potent.	$1.66 \cdot 10^{-4}$	3.78	$2.5 \cdot 10^{-28}$	27.60	[10]	—
HgJ_4^{2-}	25	0.5	"	$5.9 \cdot 10^{-3}$	2.23	$1.48 \cdot 10^{-30}$	29.83	[10]	[2, 22, 23]
JnJ^{2+}	25	1.0	Ion exch.	0.5	0.30	0.5	0.30	[11]	[24]
Pb_2J^{3+}	25	0.3—3.6	Solub.	—	—	$2.2 \cdot 10^{-2}$	1.66	[4]	—
PbJ^{2+}	25	0.3—3.6	"	$5.05 \cdot 10^{-3}$	2.30	$5.05 \cdot 10^{-3}$	2.30	[4]	[25]
PbJ_3^-	25	0.0	"	—	—	$2.22 \cdot 10^{-5}$	4.65	[12]	[17]
PbJ_4^{2-}	25	0.0	"	6.3	—0.80	$1.42 \cdot 10^{-4}$	3.85	[12]	[26]
ZnJ^+	25	3.0	Potent.	20	—1.3	20	—4.3	[13]	—

REFERENCES

1. K. B. YATSIMIRSKII, Dokl. Akad. Nauk SSSR, 77, 819 (1951);
   Z. anorg. Chem., 25, 157 (1900).
2. A. M. GOLUB, Ukrain. khim. zh., 19, 467 (1953).
3. A. K. BABKO and A. M. GOLUB, Collected Articles on General Chemistry (Sbornik statei po obshchei khimii),
   Izd. Akad. Nauk SSSR, I, 64 (1953).
4. K. B. YATSIMIRSKII and A. A. SHUTOV, Zh. fiz. khim., 27,
   782 (1953).
5. R. G. BATES and W. C. VOSBURGH, J. Am. Chem. Soc., 60,
   137 (1938).
6. J.LEDEN, Z. phys. Chem., A, 188, 160 (1941).
7. I. A. KORSHUNOV, N. I. MALYUGINA and O. M. BALABANOVA,
   Zh. obshch. khim., 21, 620 (1951).
8. G. BODLANDER and O. STORBECK, Z. anorg. Chem., 31, 458
   (1902).
9. K. B. YATSIMIRSKII and A. A. SHUTOV, Zh. fiz. khim., 26,
   842 (1952).
10. J. QUARFORT and L. G. SILLEN, Acta Chem. Scand., 3, 505
    (1949).
11. J. A. SHUFFE and H. M. EILAND, J. Am. Chem. Soc., 76,
    960 (1954).
12. E. LANFORD, ibid., 63, 667 (1941).
13. L. G. SILLEN and B. LILJEQVIST, Chem. Abs., 40, 4588
    (1946).
14. W. ERBER, Z. anorg. Chem., 248, 36 (1941);   Chem. Abs.,
    1943 (1947).
15. K. SCHULZ and B. TEZAK, Arhiv. Kemi, 23, 200 (1951).
16. G. BODLANDER and W. EBERLEIN, Z. anorg. Chem., 39, 197
    (1904).
17. I. M. KORENMAN, Zh. obshch. khim., 16, 157 (1946).
18. H. L. RILEY and B. GALLAFENT, J. Chem. Soc., 514 (1931).
19. A. T. STROMBERG and I. E. BYKOV, Zh. obshch. khim., 19,
    245 (1949).
20. P. JOB, Zbl., I, 2572 (1928).
21. H. MORZE, Z. phys. Chem., 41, 709 (1902).
22. M. S. SCHERILL, ibid., 43, 475 (1903); 47, 103 (1904).
23. N. I. MALYUGINA, M. K. SHCHENNIKOVA and I. A. KORSHUNOV,
    Zh. obshch. khim., 16, 1573 (1946).
24. N. SUNDEN, Svensk. Kem. Tidsk., 66, 50 (1954); Ref.
    Zh. khim., No. 10, 18547 (1955).
25. H. FROMHERZ and K. LIN, Z. phys. Chem., A, 153, 321
    (1931).
26. I. A. KORSHUNOV and V. A. OSIPOVA, Zh. obshch. khim.,
    19, 1816 (1949).

Nitrate Complexes

Complex ion	Temp. °C	Ionic strength	Method	k	pk	K	pK	References principal	supplementary
$BaNO_3^+$	25	0	El.cond.	0.12	0.92	0.12	0.92	[1]	—
$BiNO_3^{2+}$	25	0.1	Solub.	$5.6 \cdot 10^{-3}$	2.25	$5.6 \cdot 10^{-3}$	2.25	[2]	—
$CaNO_3^+$	25	0	El.cond.	0.52	0.28	0.52	0.28	[1]	—
$CdNO_3^+$	25	0	.	0.40	0.40	0.40	0.40	[1]	—
.	25	3.0	Potent.	0.77	0.11	0.77	0.11	[3]	—
$MgNO_3^+$	25	0	El.cond.	1.00	0	1.00	0	[1,13]	—
$PbNO_3^+$	25	2.0	Polarog.	0.33	0.48	0.33	0.48	[4]	[10]
$PuNO_3^{3+}$	25	2.0	Potent. Spectr.	0.34	0.47	0.34	0.47	[5]	[11]
$SrNO_3^+$	25	0	El.cond.	0.15	0.82	0.15	0.82	[1]	—
$ThNO_3^{3+}$	25	0.5	Distrib.	0.21	0.68	0.21	0.68	[6]	—
.	25	5.97	.	0.35	0.45	0.35	0.45	[7]	—
$Th(NO_3)_2^{2+}$	25	5.97	.	2.00	—0.30	0.71	0.15	[7]	—
$TlNO_3^{2+}$	18	3.5	Spectr.	0.66	0.18	0.66	0.18	[8]	—
$UO_2NO_3^+$	25	2.0	.	4.8	—0.68	4.8	—0.68	[9]	[12]

REFERENCES

1. E. C. RIGHELATO and C. W. DAVIES, Trans. Farad. Soc.,
   26, 592 (1930).
2. K. B. YATSIMIRSKII, Collected Articles on General Chemistry (Sbornik statei po obshchei khimii), Izd. Akad.
   Nauk SSSR, I, 97 (1953); D. F. SWINEHART and A. B.
   GARRETT, J. Am. Chem. Soc., 73, 507 (1951).
3. J. LEDEN, Z. phys. Chem., A, 188, 160 (1941).
4. H. M. HERSHENSON, M. E. SMITH and D. N. HUME, J. Am.Chem.
   Soc., 75, 507 (1953).
5. J. C. HINDMAN, Chem. Abs., 44, 3831 (1950).
6. R. A. DAY and R. W. STOUGHTON, J. Am. Chem. Soc., 72,
   5662 (1950).
7. E. L. ZEBROSKI, H. W. ALTER and F. K. NEUMANN, ibid., 73,
   5646 (1951).
8. D. PESCHANSKI, Compt. rend., 238, 2077 (1954).
9. R. BETTS and R. C. MICHELS, Chem. Abs., 44, 8745 (1950).
10. G. H. NANCOLLAS, J. Chem. Soc., 1458 (1955).
11. S. W. RABIDEAU and J. F. LAMONS, J. Am. Chem. Soc., 73,
    2895 (1951).
12. S. AHRLAND, Acta Chem. Scand., 5, 1271 (1951); Chem.
    Abs., 47, 1528 (1953).
13. H. W. JONES, C. B. MONK and C. W. DAVIES, J. Chem. Soc.,
    2693 (1949).

Selenocyanate Complexes

Complex ion	Temp. °C	Ionic Strength	Method	k	pk	K	pK
$Ag(SeCN)_2^-$	25	0.3	Potent.	—	—	$2.2 \cdot 10^{-12}$	11.66
$Ag(SeCN)_3^{2-}$	25	0.3	"	$6.0 \cdot 10^{-2}$	1.22	$1.32 \cdot 10^{-13}$	12.88
$Cd(SeCN)_4^{2-}$	25	0.3	Polaróg.	—	—	$2.5 \cdot 10^{-4}$	3.60
$Hg(SeCN)_3^-$	25	0 3	Potent.	—	—	$3.8 \cdot 10^{-27}$	26.42
$Hg(SeCN)_4^{2-}$	25	0 3	"	$3.4 \cdot 10^{-3}$	2.47	$1.29 \cdot 10^{-29}$	28.89

REFERENCE

1. V. F. TOROPOVA, Zh. neorg. khim., 1, 243 (1956).

Phosphate Complexes

Complex ion	Temp. °C	Ionic strength	Method	k	pk	K	pK	References
CaHPO$_4$	Room	0	Potent.	$4.0 \cdot 10^{-3}$	2.40	$4.0 \cdot 10^{-3}$	2.40	[1]
FeHPO$_4$$^+$	30	0.665	Spectr.	$3.5 \cdot 10^{-10}$	9.35	$3.5 \cdot 10^{-10}$	9.35	[2]
MgHPO$_4$	Room	0	Potent.	$3.2 \cdot 10^{-3}$	2.50	$3.2 \cdot 10^{-3}$	2.50	[1]
ThH$_3$PO$_4$$^{4+}$	25	2.0	Distrib.	$1.29 \cdot 10^{-2}$	1.89	$1.29 \cdot 10^{-2}$	1.89	[3]
ThH$_3$PO$_4$$^{3+}$	25	2.0	,,	$5.0 \cdot 10^{-5}$	4.30	$5.0 \cdot 10^{-5}$	4.30	[3]
Th(H$_3$PO$_4$)$_3$$^{2+}$	25	2.0	,,	$1.41 \cdot 10^{-4}$	3.85	$7.1 \cdot 10^{-9}$	8.15	[3]

REFERENCES

1. I. GREENWALD, J. REDISH and A. C. KIBRICK, J. Biol. Chem., 135, 65 (1940); Chem. Abs., 34, 7703 (1940).

2. O. E. LANFORD and S. J. KIEHL, J. Am. Chem. Soc., 64, 291 (1942); Chem. Abs., 36, 1836 (1942).

3. E. L. ZEBROSKI, H. W. ALTER and F. K. NEUMANN, J. Am. Chem. Soc., 73, 5646 (1951).

Pyrophosphate Complexes

Complex ion	Temp. °C	Ionic strength	Method	k	pk	$K.$	pK	References principal	References supplementary
$CaP_2O_7^{2-}$	19	0.02	Ind.pH meas.	$1.00 \cdot 10^{-5}$	5.00	$1.00 \cdot 10^{-5}$	5.00	[1]	[8]
$CdP_2O_7^{2-}$	—	—	Polarog.	—	—	$2.7 \cdot 10^{-6}$	5.57	[2]	[9]
$CuP_2O_7^{2-}$	25	0.1	Solub.	$2.0 \cdot 10^{-7}$	6.70	$2.0 \cdot 10^{-7}$	6.70	[3]	[10,11]
$Cu(P_2O_7)_2^{6-}$	25	0.1	"	$5 \cdot 10^{-3}$	2.30	$1.0 \cdot 10^{-9}$	9.00	[3]	[11,12]
$MgP_2O_7^{2-}$	19	0.02	Ind.pH meas.	$2.0 \cdot 10^{-6}$	5.70	$2.0 \cdot 10^{-6}$	5.70	[4]	—
$NiP_2O_7^{2-}$	25	0.1	Solub.	$1.5 \cdot 10^{-6}$	5.82	$1.5 \cdot 10^{-6}$	5.82	[3]	—
$Ni(P_2O_7)_2^{6-}$	25	0.1	"	$4.3 \cdot 10^{-2}$	1.37	$6.5 \cdot 10^{-8}$	7.19	[3]	—
$Pb(P_2O_7)_2^{6-}$	25	—	El.cond.	—	—	$4.74 \cdot 10^{-2}$	5.33	[5]	[13]
$TlP_2O_7^{3-}$	35	—	Polarog.	$2.0 \cdot 10^{-2}$	1.69	$2.0 \cdot 10^{-2}$	1.69	[6]	—
$Tl(P_2O_7)_2^{7-}$	35	—	Polarog.	0.66	0.18	$1.35 \cdot 10^{-2}$	1.87	[6]	—
$Zn(P_2O_7)_2^{6-}$	25	—	Cryosc., Polarog.	—	—	$3.4 \cdot 10^{-7}$	6.46	[2,7]	[12]
$CeP_2O_7^{-}$	25	0.5	Ion exch.	$7.1 \cdot 10^{-18}$	17.15	$7.1 \cdot 10^{-18}$	17.15	[14]	—

REFERENCES

1. K. B. YATSIMIRSKII and V. P. VASIL'EV, Zh. fiz. khim.,
   30, 28 (1956).
2. G. SARTORY, Gazz. chim. Ital., 64, 3 (1934).
3. K. B. YATSIMIRSKII and V. P. VASIL'EV, Zh. anal. khim.,
   11, 536 (1956).
4. V. P. VASIL'EV, Zh. fiz. khim. (in the press).
5. B. C. HALDAR, Current Sci., 19, 244 (1950); Zbl., I,
   2856 (1951).
6. P. SENISE and P. DELAHAY, J. Am. Chem. Soc., 74, 6128
   (1952).
7. B. C. HALDAR, Current Sci., 19, 283 (1950); Chem. Abs.,
   45, 9415 (1951).
8. R. GOSSELIN and E. COGHLAN, Arch. Biochem. Biophys., 45,
   301 (1953).
9. P. SOUCHEY and J. FENCHERRE, Bull. Soc. chim. France,
   529 (1947); Chem. Abs., 42, 810 (1948).
10. E. ERIKSON, Chem. Abs., 44, 3392 (1950).
11. J. WATTERS and A. A. AARON, J. Am. Chem. Soc., 75, 611
    (1953).
12. A. I. STABROVSKII, Zh. fiz. khim., 26, 949 (1952).
13. K. B. YATSIMIRSKII and V. P. VASIL'EV, ibid., 30, 901
    (1956).
14. S. W. MAYER and S. D. SCHWARTZ, J. Am. Chem. Soc., 72,
    5106 (1950).

Perchlorate Complexes

Complex ion	Temp. °C	Ionic strength	Method	k	pk	K	pK	Ref.
CeClO$_4^{3+}$	25	0	Spectr.	$1.17 \cdot 10^{-2}$	1 93	$1.17 \cdot 10^{-2}$	1.93	[1]
	25	4.5	Spectr.	1.16	−0.07	1.16	−0.07	[1]
FeClO$_4^{2+}$	25	0	Spectr.	2.10	−0.3	2.10	−0,32	[2]

REFERENCES

1. L. J. HEIDT and J. BERESTECKI, J. Am. Chem. Soc., 77,
   2049 (1955).
2. J. SUTTON, Nature, 169, 71 (1952).

Sulphate Complexes

Complex ion	Temp. °C	Ionic strength	Method	k	pk	K	pK	References principal	References supplementary
$AgSO_4^-$	25	3.0	Potent.	0.59	0.23	0.59	0.23	[1]	—
$Ag(SO_4)_2^{3-}$	25	3.0	"	1.0	0.00	0.59	0.23	[1]	—
$CaSO_4$	25	0	Solub.	$4.9 \cdot 10^{-3}$	2.31	$4.9 \cdot 10^{-3}$	2.31	[2]	[3]
$CdSO_4$	25	0	El.cond.	$4.9 \cdot 10^{-3}$	2.31	$4.9 \cdot 10^{-3}$	2.31	[3]	—
$CeSO_4^+$	25	3.0	Potent.	$1.41 \cdot 10^{-2}$	0.85	$1.41 \cdot 10^{-2}$	0.85	[1]	[25]
"	25	0	Solub., Spectr.	$4.0 \cdot 10^{-4}$	3.40	$4.0 \cdot 10^{-4}$	3.40	[4]	[9,25]
$CoSO_4$	25	0	Ion exch.	$1.66 \cdot 10^{-2}$	1.78	$1.66 \cdot 10^{-2}$	1.78	[5]	—
$CuSO_4$	25	0	El.cond.	$3.4 \cdot 10^{-3}$	2.47	$3.4 \cdot 10^{-3}$	2.47	[3]	—
"	20	1.0	Potent.	$4.5 \cdot 10^{-3}$	2.35	$4.5 \cdot 10^{-3}$	2.35	[6]	—
"	25	4.0	Spectr.	0.112	0.95	0.112	0.95	[7]	—
$ErSO_4^+$	25	0	El.cond.	0.42	0.38	0.42	0.38	[8]	—
$FeSO_4$	25	0	Calc.	$2.6 \cdot 10^{-4}$	3.58	$2.6 \cdot 10^{-4}$	3.58	[9]	—
$FeSO_4^+$	18	0	Kin.	$5 \cdot 10^{-3}$	2.30	$5 \cdot 10^{-3}$	2.30	[10]	—
"	18	0.066	Ion exch.	$6.8 \cdot 10^{-5}$	4.17	$6.8 \cdot 10^{-5}$	4.17	[11]	—
"			Spectr.	$1.0 \cdot 10^{-3}$	3.00	$1.0 \cdot 10^{-3}$	3.00	[11]	—
$Fe(SO_4)_2^-$	28	1.0	"	$9.3 \cdot 10^{-3}$	2.03	$9.3 \cdot 10^{-3}$	2.03	[12]	—
$GdSO_4^+$	28	1.0	"	0.112	0.95	$1.05 \cdot 10^{-3}$	2.98	[12]	—
HSO_4^-	25	0	El.cond.	$2.2 \cdot 10^{-4}$	3.66	$2.2 \cdot 10^{-4}$	3.66	[9]	—
$HoSO_4^+$	25	0	"	$7.6 \cdot 10^{-2}$	1.12	$7.6 \cdot 10^{-2}$	1.12	[13]	—
$InSO_4^+$	25	0	"	$2.6 \cdot 10^{-4}$	3.58	$2.6 \cdot 10^{-4}$	3.58	[9]	—
"	20	2.0	Potent.	$1.66 \cdot 10^{-2}$	1.78	$1.66 \cdot 10^{-2}$	1.78	[14]	[27]
$In(SO_4)_2^-$	20	2.0	"	0.79	0.10	$1.32 \cdot 10^{-2}$	1.88	[14]	—
$In(SO_4)_3^{3-}$	20	2.0	"	0.32	0.48	$4.4 \cdot 10^{-3}$	2.36	[14]	—
KSO_4^-	25	0	El.cond.	0.11	0.96	0.11	0.96	[9]	—

Sulphate Complexes (contd.)

Complex ion	Temp. °C	Ionic strength	Method	k	p^k	K	pK	References principal	References supplementary
$LaSO_4^+$	25	0	Solub.	$2.3 \cdot 10^{-4}$	3.64	$2.3 \cdot 10^{-4}$	3.64	[15]	[9,13,16]
$MgSO_4$	25	0	Potent.	$4.4 \cdot 10^{-3}$	2.35	$4.4 \cdot 10^{-3}$	2.35	[16]	[3]
$MnSO_4$	25	0	El.cond.	$5.2 \cdot 10^{-3}$	2.28	$5.2 \cdot 10^{-3}$	2.28	[10]	—
$NaSO_4^-$	25	0	"	0.19	0.72	0.19	0.72	[13]	—
$NdSO_4^+$	25	0	"	$2.3 \cdot 10^{-4}$	3.64	$2.3 \cdot 10^{-4}$	3.64	[9]	—
$NiSO_4$	25	0	Distrib.	$4.0 \cdot 10^{-3}$	2.40	$4.0 \cdot 10^{-3}$	2.40	[3]	—
$NpSO_4^{2+}$	25	2.0	"	$3.7 \cdot 10^{-3}$	2.43	$3.7 \cdot 10^{-3}$	2.43	[17]	—
$Np(SO_4)_2$	25	2.0	El.cond.	$9.1 \cdot 10^{-2}$	1.04	$3.4 \cdot 10^{-4}$	3.47	[17]	—
$PrSO_4^+$	25	2.0	Potent.	$2.4 \cdot 10^{-4}$	3.62	$2.4 \cdot 10^{-4}$	3.62	[9]	—
$PuSO_4^{2+}$	25	1.0	El.cond.	$2.2 \cdot 10^{-4}$	3.66	$2.2 \cdot 10^{-4}$	3.66	[18]	—
$SmSO_4^+$	25	0	Distrib.	$2.2 \cdot 10^{-4}$	3.66	$2.2 \cdot 10^{-4}$	3.66	[9]	—
$ThSO_4^{2+}$	25	2.0	"	$4.8 \cdot 10^{-4}$	3.32	$4.8 \cdot 10^{-4}$	3.32	[19,20]	—
$Th(SO_4)_3$	25	2.0	"	$6.6 \cdot 10^{-3}$	2.18	$3.2 \cdot 10^{-6}$	5.50	[19,20]	—
USO_4^{2+}	25	3.5	"	$5.8 \cdot 10^{-4}$	3.24	$5.8 \cdot 10^{-4}$	3.24	[21,22]	—
$U(SO_4)_2$	25	3.5	"	$6.6 \cdot 10^{-3}$	2.18	$3.8 \cdot 10^{-6}$	5.42	[21,22]	—
UO_2SO_4	20	1.0	Potent.	0.178	1.70	$2.0 \cdot 10^{-2}$	1.70	[23]	[29]
$UO_2(SO_4)_2^{2-}$	20	1.0	"	0.141	0.75	$2.8 \cdot 10^{-3}$	2.55	[23]	—
$UO_2(SO_4)_3^{4-}$	20	1.0	"		0.85	$4.0 \cdot 10^{-4}$	3.40	[23]	—
YSO_4^+	25	0	El.cond.	$3.4 \cdot 10^{-4}$	3.47	$3.4 \cdot 10^{-4}$	3.47	[9]	—
$YbSO_4^+$	25	0	"	$2.6 \cdot 10^{-4}$	3.58	$2.6 \cdot 10^{-4}$	3.58	[9]	—
$ZnSO_4$	25	0	"	$4.9 \cdot 10^{-3}$	2.31	$4.9 \cdot 10^{-3}$	2.31	[6]	—
$ZrSO_4^{2+}$	25	2.0	Distrib.	$1.62 \cdot 10^{-4}$	3.79	$1.62 \cdot 10^{-4}$	3.79	[24]	—
$Zr(SO_4)_2$	25	2.0	"	$1.41\ 10^{-3}$	2.85	$2.3 \cdot 10^{-7}$	6.64	[24]	—
$Zr(SO_4)_3^{2-}$	25	2.0	"	$7.4 \cdot 10^{-2}$	1.13	$1.7 \cdot 10^{-8}$	7.77	[24]	—

REFERENCES

1. I.LEDEN, Acta Chem.Scand., 6, 97 (1952); Chem.Abs., 47, 3747 (1953).
2. R.P.BELL and J.H.B.GEORGE, Trans.Farad.Soc., 49, 619 (1953).
3. T.O.DENNEY and C.B.MONK, ibid., 47, 992 (1951).
4. T.W.NEWTON and G.M.ARCAND, J.Am.Chem.Soc., 75, 2449 (1953).
5. R.E.CONNICK and S.W.MAYERS, ibid., 73, 1176 (1951).
6. B.B.OWEN and R.W.GURRY, ibid., 60, 3074 (1938).
7. S.FRONAEUS, Acta Chem.Scand., 4, 72 (1950); Chem.Abs., 44, 8810 (1950).
8. R.NASANEN, Suomen Kemistileht, 67, 2613 (1953); Chem. Abs., 48, 11158 (1954).
9. F.H.SPEDDING and S.JAFFE, J.Am.Chem.Soc., 76, 882 (1954).
10. K.B.YATSIMIRSKII, Zh.obshch.khim., 24, 1498 (1954).
11. K.W.SYKES, J.Chem.Soc., 124 (1952).
12. R.A.WHITEKER and N.DAVIDSON, J.Am.Chem.Soc., 75, 3081 (1953).
13. J.L.JENKINS and C.B.MONK, ibid., 72, 2695 (1950).
14. N. SUNDEN, Svensk.Kem.Tidsk., 65, 257 (1953); Ref. zh. khim., 18546 (1955).
15. C.W.DAVIES, J.Chem.Soc., 2421 (1930).
16. H.W.JONES and C.B.MONK, Trans.Farad.Soc., 48, 929 (1952).
17. J.C.SULLIVAN and J.C.HINDMAN, J.Am.Chem.Soc., 76, 5931 (1954).
18. S.W.RABIDEAU and J.F.LAMONS, ibid., 73, 2895 (1951).
19. R.A.WHITEKER and N.DAVIDSON, ibid., 75, 3081 (1953).
20. E.L.ZEBROSKI, H.W.ALTER and F.K.NEUMANN, ibid., 73, 5646 (1951).
21. R.H.BETTS and R.LEIGH, Canad.J.Res., B8, 514 (1950).
22. J.C.SULLIVAN and J.C.HINDMAN, J.Am.Chem.Soc., 74, 6091 (1952).
23. S.AHRLAND, Acta Chem.Scand., 5, 1151 (1951); Chem.Abs., 46, 5480 (1952).
24. R.E.CONNICK and W.H.McVEY, J.Am.Chem.Soc., 71, 3182 (1949).
25. I.LEDEN, Z.phys.Chem., A, 188, 160 (1941); Chem. Abs., 48, 9251 (1954).
26. I.M.KORENMAN, Zh.obshch.khim., 24, 1910 (1954).
27. N.SUNDEN, Svensk.Kem.Tidsk., 66, 173 (1954); Chem.Abs., 49, 1465 (1955).
28. R.A.DAY, R.N.WILHITE and F.D.HAMILTON, J.Am.Chem.Soc., 77, 3180 (1955).
29. R.H.BETTS and R.K.MICHELS, Chem.Abs., 44, 8745 (1950).

Sulphite Complexes

Complex ion	Temp. °C	Ionic strength	Method	k	p_k	K	p_K	References principal	References supplementary
$AgSO_3^-$	25	2.0	Potent.	$5.0 \cdot 10^{-6}$	5.30	$5.0 \cdot 10^{-6}$	5.30	[1]	—
$Ag(SO_3)_2^{3-}$	25	2.0	"	$8.9 \cdot 10^{-3}$	2.05	$4.5 \cdot 10^{-8}$	7.35	—	[3]
$CuSO_3^-$	20	1.0	Polarog.	$3.4 \cdot 10^{-8}$	7.47	$3.4 \cdot 10^{-8}$	7.47	[1]	—
$Cu(SO_3)_2^{3-}$	20	1.0	"	$9.1 \cdot 10^{-2}$	1.04	$3.1 \cdot 10^{-9}$	8.51	[1]	—
$Cu(SO_3)_3^{5-}$	20	1.0	"	0.209	0.68	$6.5 \cdot 10^{-10}$	9.19	[1]	—
$Hg(SO_3)_2^{2-}$	18	3.0	Potent.	—	—	$2.19 \cdot 10^{-23}$	22.66	[2]	—
$Hg(SO_3)_3^{4-}$	18	3.0	"	0.78	0.11	$1.70 \cdot 10^{-23}$	22.77	[2]	—
$Hg(SO_3)_4^{6-}$	18	3.0	"	0.85	0.07	$1.45 \cdot 10^{-23}$	22.84	[2]	—

REFERENCES

1. V. F. TOROPOVA, I. A. SIROTINA and V. B. ROTANOVA, Uch. zap. Kazansk. Gosuniv. im. Ul'yanova-Lenina, 115, No. 3, 53 (1955).
2. V. F. TOROPOVA, and E. A. BYELAYA, ibid., 115, No. 3, 61 (1955).
3. E. L. JAHN and H. STANDE, Z. Naturforsch., 6A, 385 (1951).

Tetrametaphosphate Complexes

Complex ion	Temp. °C	Ionic strength	Method	κ	$p\kappa$	K	pK	References
$BaP_4O_{12}^{2-}$	25	0	El.cond.	$1.02 \cdot 10^{-5}$	4.99	$1.02 \cdot 10^{-5}$	4.99	[1]
$CaP_4O_{12}^{2-}$	25	0	"	$3.4 \cdot 10^{-6}$	5.47	$3.4 \cdot 10^{-6}$	5.47	[1]
$CuP_4O_{12}^{2-}$	30	1.0	Potent.	$7.6 \cdot 10^{-4}$	3.12	$7.6 \cdot 10^{-4}$	3.12	[2]
$Cu(P_4O_{12})_2^{6-}$	30	1.0	"	$3.0 \cdot 10^{-2}$	1.52	$2.3 \cdot 10^{-5}$	4.64	[2]
$LaP_4O_{12}^{-}$	25	0	El.cond.	$2.2 \cdot 10^{-7}$	6.66	$2.2 \cdot 10^{-7}$	6.66	[3]
$MgP_4O_{12}^{2-}$	25	0	"	$6.8 \cdot 10^{-6}$	5.17	$6.8 \cdot 10^{-6}$	5.17	[1]
$MnP_4O_{12}^{2-}$	25	0	"	$3.3 \cdot 10^{-6}$	5.48	$3.3 \cdot 10^{-6}$	5.48	[1]
$NaP_4O_{12}^{3-}$	30	0	Potent.	0.15	0.81	0.15	0.81	[2]
$NiP_4O_{12}^{2-}$	25	0	El.cond.	$1.12 \cdot 10^{-5}$	4.95	$1.12 \cdot 10^{-5}$	4.95	[1]
$NiP_4O_{12}^{2-}$	30	1.0	Potent.	$2.3 \cdot 10^{-3}$	2.63	$2.3 \cdot 10^{-3}$	2.63	[2]
$Ni(P_4O_{12})_2^{6-}$	30	1.0	"	0.14	0.85	$3.3 \cdot 10^{-4}$	3.48	[2]
$SrP_4O_{12}^{2-}$	25	0	El.cond., Solub.	$7.8 \cdot 10^{-6}$	5.11	$7.8 \cdot 10^{-6}$	5.11	[4]

REFERENCES

1. H. W. JONES and C. B. MONK, J. Chem. Soc., 3475 (1950).

2. R. J. GROSS and J. W. GRYDER, J. Am. Chem. Soc., 77, 3695 (1955).

3. C. B. MONK, J. Chem. Soc., 1317 (1952).

4. C. B. MONK, ibid., 1314 (1952).

Thiocyanate Complexes

Complex ion	Temp. °C	Ionic strength	Method	k	pk	K	pK	References principal	References supplementary
$Ag(CNS)_2^-$	25	2.2	Solub.	—	—	$2.7 \cdot 10^{-8}$	7.57	[1]	[22, 23]
$Ag(CNS)_3^{2-}$	25	2.2	"	$3.1 \cdot 10^{-3}$	2.51	$8.3 \cdot 10^{-10}$	9.08	[1]	[24] [22, 24, 25]
$Ag(CNS)_4^{3-}$	25	2.2	"	0.10	1.00	$9.3 \cdot 10^{-11}$	10.08	[1]	[24, 25]
$Au(CNS)_2^-$	Room	—	Potent.	—	—	10^{-23}	23	[2]	—
$Au(CNS)_4^-$	.	—	"	—	—	10^{-42}	42	[2]	—
$BiCNS^{++}$	20—25	Variab	Spectr.	$7.15 \cdot 10^{-2}$	1.15	$7.15 \cdot 10^{-2}$	1.15	[3]	—
$Bi(CNS)_2^+$	20—25	"	"	$7.7 \cdot 10^{-2}$	1.11	$5.5 \cdot 10^{-3}$	2.26	[3]	[26]
$Bi(CNS)_4^-$	20—25	"	"	—	—	$3.93 \cdot 10^{-4}$	3.41	[3]	—
$Bi(CNS)_6^{3+}$	20—25	"	"	—	—	$5.93 \cdot 10^{-5}$	4.23	[3]	[22]
$CdCNS^+$	30	2.0	Polarog.	$9 \cdot 1 \cdot 10^{-2}$	1.04	$9 \cdot 1 \cdot 10^{-2}$	1.04	[4]	—
"	25	3.0	Potent.	$4 \cdot 1 \cdot 10^{-2}$	1.39	$4 \cdot 1 \cdot 10^{-2}$	1.39	[5]	—
$Cd(CNS)_2$	30	2.0	Polarog.	0.195	0.71	$1.8 \cdot 10^{-2}$	1.75	[4]	—
"	25	3.0	Potent.	0.256	0.59	$1.05 \cdot 10^{-2}$	1.98	[5]	—
$Cd(CNS)_3^-$	30	2.0	Polarog.	(9.3)	(−0.97)	(0.167)	(0.78)	[4]	[6]
"	25	3.0	Potent.	0.25	0.60	$2.6 \cdot 10^{-3}$	2.58	[5]	—
$Cd(CNS)_4^{2-}$	30	2.0	Polarog.	0.10	1.00	$1.67 \cdot 10^{-2}$	1.78	[4]	[6]

Thiocyanate Complexes (contd.)

Complex ion	Temp. °C	Ionic strength	Method	k	p_k	K	pK	References principal	References supplementary
Cd (CNS)$_6^{4-}$	25	0.5—5.0	Polarog.	—	—	(1.033)	(—0.08)	[6]	—
Co CNS$^+$	25	—	Spectr.	1.08	—0.04	1.08	—0.04	[7]	—
Co (CNS)$_2$	25	—	"	4.6	—0.66	5 0	—0.70	[7]	—
Co (CNS)$_3^-$	25	—	"	0.2	0.70	1.00	0.00	[7]	—
Co (CNS)$_4^{2-}$	25	—	Analysis	$5 \cdot 10^{-3}$	2.30	$1.00 \cdot 10^{-3}$	3.00	[7]	—
CrCNS^{2+}	25	1.0	"	$1.35 \cdot 10^{-2}$	1.87	$1.35 \cdot 10^{-2}$	1.87	[8]	[27]
Cr (CNS)$_2^+$	25	1.0	"	$7.7 \cdot 10^{-2}$	1.11	$1.05 \cdot 10^{-3}$	2.98	[8]	—
Cu (CNS)$_2^-$	18	3.09	Potent.	—	—	$7.83 \cdot 10^{-13}$	12.11	[9]	—
Cu (CNS)$_3^-$	18	4.2	Spectr.	—	—	$6.5 \cdot 10^{-6}$	5.18	[10]	—
FeCNS^{2+}	25	0.0	"	$1.12 \cdot 10^{-3}$	2.95	$1.12 \cdot 10^{-3}$	2.95	[11]	[28,29,30,31]
Fe (CNS)$_2^+$	25	1.28	"	$8.7 \cdot 10^{-3}$	2.06	$8.7 \cdot 10^{-3}$	2.06	[12]	—
Fe (CNS)$_3$	25	1.28	"	$5.0 \cdot 10^{-2}$	1.30	$4.4 \cdot 10^{-4}$	3.36	[12]	—
Hg (CNS)$_2$	25	0.35	"	—	—	$3.4 \cdot 10^{-18}$	17.47	[13]	—
Hg (CNS)$_4^{2-}$	25	0.3	Potent.	—	—	$5.9 \cdot 10^{-22}$	21.23	[14]	[32,33]
InCNS^{2+}	20	2.0	"	$2.6 \cdot 10^{-3}$	2.58	$2.6 \cdot 10^{-3}$	2.58	[15]	—
In (CNS)$_2^+$	20	2.0	"	0.38	0.42	$1.0 \cdot 10^{-3}$	3.00	[15]	—
In (CNS)$_3$	20	2.0	"	$2.3 \cdot 10^{-2}$	1.63	$2.3 \cdot 10^{-5}$	4.63	[15]	—
NiCNS$^+$	20	1.5	Ion exch.	$6.7 \cdot 10^{-2}$	1.18	$6.7 \cdot 10^{-2}$	1.18	[16]	—
Ni (CNS)$_2$	20	1.5	"	0.35	0.46	$2.3 \cdot 10^{-2}$	1.64	[16]	—

Thiocyanate Complexes (contd.)

Complex ion	Temp. °C	Ionic strength	Method	k	pk	K	pK	References principal	supplementary
Ni(CNS)$_3^-$	20	1.5	Ion exch.	0.68	0.17	$1.55 \cdot 10^{-2}$	1.81	[16]	—
Pb(CNS)$_6^{4-}$	25	3.0—7.5	Solub.	—	—	2.0	−0.30	[17]	—
RuCNS^{2+}	25	1.0	Spectr.	$1.17 \cdot 10^{-2}$	1.78	$1.17 \cdot 10^{-2}$	1.78	[18]	—
TlCNS	25	0.0	Solub.	0.160	0.80	0.160	0.80	[19]	—
UCNS^{3+}	25	2.00	Potent.	$3.2 \cdot 10^{-2}$	1.49	$3.2 \cdot 10^{-2}$	1.49	[20]	—
U(CNS)$_2^{2+}$	25	2.00	"	0.24	0.62	$7.7 \cdot 10^{-3}$	2.11	[20]	—
UO$_2$CNS$^+$	20	0.1	Spectr.	0.174	0.76	0.174	0.76	[21]	—
UO$_2$(CNS)$_2$	20	0.1	"	1.05	−0.02	0.182	0.74	[21]	—
UO$_2$(CNS)$_3^-$	20	0.1	"	0.36	0.44	$6.6 \cdot 10^{-2}$	1.18	[21]	—
VCNS^{2+}	25	2.6	"	10^{-2}	2.0	10^{-2}	2.0	[11]	—
VOCNS$^+$	25	2.6	"	0.12	0.92	0.12	0.92	[11]	—
ZnCNS$^+$	18	0.1	Potent.	$2.4 \cdot 10^{-2}$	1.62	$2.4 \cdot 10^{-2}$	1.62	[22]	—

REFERENCES

1. G. S. CAVE and N. D. HUME, J. Am. Chem. Soc., 75, 2893 (1953).

2. N. BJERRUM and A. KIRSCHNER, Kgl. Danske Videnskab. Math. Phys., V, No. 1 (1918).

3. W. D. KINGERY and D. N. HUME, J. Am. Chem. Soc., 71, 2393 (1949).

4. D. N. HUME, D. D. De FORD and G. S. CAVE, ibid., 73, 5323 (1951).

5. J. LEDEN, Z. phys. Chem., A, 188, 160 (1941).

6. I. A. KORSHUNOV, N. I. MALYUGINA, O. M. BALABANOVA, Zh. obshch. khim., 21, 620 (1951).

7. M. LEHNE, Bull. Soc. chim. France, 76 (1951); Chem. Abs., 45, 6117 (1951).

8. K. G. POULSEN, J. BJERRUM and J. POULSEN, Acta Chem. Scand., 8, 921 (1954); Chem. Abs., 49, 2926 (1955).
9. A. I. STABROVSKII, Zh. fiz. khim., 26, 949 (1952).
10. M. OUDINET and F. GALLAIS, Compt. rend., 373 (1953); Chem. Abs., 49, 13816 (1955).
11. S. C. FURMAN and S. C. GARNER, J. Am. Chem. Soc., 73, 4528 (1951).
12. R. H. BETTS and F. S. DAINTON, ibid., 75, 5721 (1953).
13. K. B. YATSIMIRSKII and B. D. TUKHLOV, Zh. obshch. khim., 26, 356 (1956).
14. V. F. TOROPOVA, Zh. neorg. khim., 1, 243 (1956).
15. N. SUNDEN, Svensk. Kem. Tidsk., 66, 50 (1954); Ref. zh. khim., 18547 (1955).
16. S. FRONAEUS, Acta Chem. Scand., 7, 21 (1953); Chem. Abs., 47, 8582 (1953).
17. K. B. YATSIMIRSKII, Zh. fiz. khim., 25, 475 (1951).
18. R. P. VAFFE and A. VOIGT, J. Am. Chem. Soc., 74, 2500 (1952).
19. R. P. BELL and J. H. B. GEORGE, Trans. Farad. Soc., 49, 619 (1953).
20. R. A. DAY, R. N. WILHITE and F. D. HAMILTON, J. Am. Chem. Soc., 77, 3180 (1955).
21. S. AHRLAND, Chem. Abs., 44, 5256 (1950).
22. E. FERRELL, J. M. RIDGION and H. L. RILEY, J. Chem. Soc., 1121 (1936).
23. G. BODLANDER and W. EBERLEIN, Z. anorg. Chem., 39, 197 (1904).
24. J. LEDEN, Z. Naturforsch., No. 1, 10a (1955).
25. J. KRATOHVIL, cited in V. B. Vonk, J. Kratohvil and B. Tezak, Arhiv. Kemi, 25, 219 (1953).
26. F. S. FRUM and M. N. SKOBINA, Ref. zh. khim., 28666 (1954).
27. N. BJERRUM, Z. anorg. Chem., 119, 189 (1921).
28. A. K. BABKO, Zh. obshch. khim., 16, 1549 (1946).
29. S. M. EDMONDS and N. BIRNBAUM, J. Am. Chem. Soc., 63, 1471 (1941).
30. H. E. BENT and C. L. FRENCH, ibid., 63, 568 (1941).
31. H. S. FRANK and R. L. OSWALT, ibid., 69, 1321 (1947).
32. N. A. KORSHUNOV and M. K. SHCHENNIKOVA, Zh. obshch. khim., 19, 1820 (1949).
33. F. GALLAIS and J. MONNIER, Compt. rend., 223, 790 (1946).
34. S. AHRLAND and R. LARSSON, Acta Chem. Scand., 8, 137 (1954); Chem. Abs., 48, 11969 (1954).

Thiosulphate Complexes

Complex ion	Temp. °C	Ionic strength	Method	k	pk	K	pK	References principal	References supplementary
$AgS_2O_3^-$	20	—	Potent.	$1.5 \cdot 10^{-9}$	8.82	$1.5 \cdot 10^{-9}$	8.82	[1]	—
$Ag(S_2O_3)_2^{3-}$	20	—	Solub.	$2.3 \cdot 10^{-5}$	4.64	$3.5 \cdot 10^{-14}$	13.46	[1]	[10]
BaS_2O_3	25	0	Solub.	$4.7 \cdot 10^{-3}$	2.33	$4.7 \cdot 10^{-3}$	2.33	[2]	[11]
CaS_2O_3	25	0	.	$1.04 \cdot 10^{-2}$	1.98	$1.04 \cdot 10^{-2}$	1.98	[2]	[11]
CdS_2O_3	25	0	.	$1.21 \cdot 10^{-4}$	3.92	$1.21 \cdot 10^{-4}$	3.92	[2]	—
.	18	0.11	Kin.	$1.26 \cdot 10^{-3}$	2.90	$1.26 \cdot 10^{-7}$	6.90	[3]	—
$Cd(S_2O_3)_2^{2-}$	25	0	.	$3.0 \cdot 10^{-3}$	2.52	$3.6 \cdot 10^{-7}$	6.44	[2]	[10]
$Cd(S_2O_3)_3^{4-}$	25	0.3—9.0	Polarog.	—	—	$4.7 \cdot 10^{-7}$	6.33	[4]	[10]
CoS_2O_3	25	0	Solub.	$9.0 \cdot 10^{-3}$	2.05	$9.0 \cdot 10^{-8}$	7.05	[2]	—
$CuS_2O_3^-$	25	2.0	Polarog., Potent.	$5.4 \cdot 10^{-11}$	10.27	$5.4 \cdot 10^{-11}$	10.27	[5]	—
$Cu(S_2O_3)_2^{3-}$	25	2.0	.	$1.12 \cdot 10^{-2}$	1.95	$6.0 \cdot 10^{-13}$	12.22	[5]	[12]
$Cu(S_2O_3)_3^{5-}$	25	2.0	.	$2.4 \cdot 10^{-2}$	1.62	$1.44 \cdot 10^{-14}$	13.84	[5]	—
FeS_2O_3	6,1	0	Kin.	$6.8 \cdot 10^{-3}$	2.17	$6.8 \cdot 10^{-3}$	2.17	[6]	—
$FeS_2O_3^+$	6,1	0.48	.	0.12	0.92	0.12	0.92	[6]	[13]
.	6,1	0	Spectr.	$5.59 \cdot 10^{-4}$	3.25	$5.59 \cdot 10^{-4}$	3.25	[6]	—
.	25	0.47	.	$2.55 \cdot 10^{-2}$	1.59	$2.55 \cdot 10^{-2}$	1.59	[6]	—
.	25	0.47	.	$7.9 \cdot 10^{-3}$	2.10	$7.9 \cdot 10^{-3}$	2.10	[6]	—
$Hg(S_2O_3)_2^{2-}$	25	0	Potent.	—	—	$3.6 \cdot 10^{-30}$	29.44	[7]	—
$Hg(S_2O_3)_3^{4-}$	25	0	.	$3.5 \cdot 10^{-3}$	2.45	$1.26 \cdot 10^{-32}$	31.90	[7]	—

Thiosulphate Complexes (contd.).

Complex ion	Temp. °C	ionic strength	Method	k	p^k	K	pK	References principal	supplementary
$Hg(S_2O_3)_4^{6-}$	25	0	Potent.	$4.6 \cdot 10^{-2}$	1.34	$2.8 \cdot 10^{-31}$	33.24	[7]	—
$KS_2O_3^{-}$	25	0	Solub.	0.12	0.92	0.12	0.92	[2]	—
MgS_2O_3	25	0	.	$1.45 \cdot 10^{-2}$	1.84	$1.45 \cdot 10^{-2}$	1.84	[2]	—
MnS_2O_3	25	0	.	$1.12 \cdot 10^{-2}$	1.95	$1.12 \cdot 10^{-2}$	1.95	[2]	—
$NaS_2O_3^{-}$	25	0	.	**0.21**	0.68	0.21	0.68	[2]	—
NiS_2O_3	25	0	.	$8.7 \cdot 10^{-3}$	2.06	$8.7 \cdot 10^{-5}$	2.06	[2]	—
$Pb(S_2O_3)_2^{2-}$	25	0.07—0.75	.	$7.41 \cdot 10^{-6}$	5 13	$7.41 \cdot 10^{-6}$	5.13	[8]	—
$Pb(S_2O_3)_3^{4-}$	25	0.07—0.75	.	$6.0 \cdot 10^{-22}$	1.22	$4.48 \cdot 10^{-7}$	6.35	[8]	—
SrS_2O_3	25	0	.	$9.2 \cdot 10^{-3}$	2.04	$9.2 \cdot 10^{-3}$	2.04	[2]	—
$TlS_2O_3^{-}$	Room	0.1—0.2	Polarog.	$1.23 \cdot 10^{-2}$	1.91	$1.23 \cdot 10^{-2}$	1.91	[9]	—
ZnS_2O_3	25	0	.	$4.0 \cdot 10^{-3}$	2.40	$4.0 \cdot 10^{-3}$	2.40	[2]	[10]

REFERENCES

1. H.CHATEAU and J.POURADIER, Compt.rend., 240, 1882 (1955).
2. T.O.DENNEY and C.B.MONK, Trans.Farad.Soc., 47, 992 (1951).
3. K.B.YATSIMIRSKII, Zh.anal.khim., 6, 344 (1955).
4. A.G.STROMBERG and -.E.BYKOV, Zh.obshch.khim., 19, 245 (1949).
5. V.F.TOROPOVA, I.A.SIROTINA and T.I.LISOVA, Uch. zap.Kazansk.Gosuniv.im.Ul'yanova-Lenina, 115, No.3, 43 (1955).
6. F.M.PAGE, Trans.Farad.Soc., 50, 120 (1954).
7. V.F.TOROPOVA, Zh.obshch.khim., 24, 423 (1954).
8. K.B.YATSIMIRSKII, Zh.fiz.khim., 25, 475 (1951).
9. M.S.NOVAKOVSKII and T.M.SHMAYEVA, Ukrain.khim.zh., 20, 615 (1954).
10. E.FERELL, J.M.RIDGION and H.L.RILEY, J.Chem.Soc., 1121 (1936).
11. C.W.DAVIES and P.A.H.WYATT, Trans. Farad.Soc., 45, 770 (1949).
12. A.I.STABROVSKII, Zh.fiz.khim., 26, 949 (1952).
13. B.C.HALDAR and S.BANERJEE, Chem.Abs., 43, 6536 (1949).

Thiourea Complexes

Complex ion	Temp. °C	Ionic strength	Method	k	pk	K	pK	References principal	supplementary
$Ag(CSN_2H_4)_3^+$	Room	0.01	Potent.	·	—	$7.0 \cdot 10^{-14}$	13.14	[1]	[7]
$Bi(CSN_2H_4)_6^{3+}$	·	1.0	Polarog.	—	—	$1.4 \cdot 10^{-12}$	11.94	[2]	—
$Cd(CSN_2H_4)^{2+}$	25	0.2	"	$2\ 63 \cdot 10^{-2}$	1.58	$2.63 \cdot 10^{-2}$	1.58	[3]	—
$Cd(CSN_2H_4)_2^{2+}$	25	0.2	"	$8.9 \cdot 10^{-2}$	1.05	$2.3 \cdot 10^{-3}$	2.63	[3]	—
$Cd(CSN_2H_4)_3^{2+}$	25	0.1	"	—	—	$1.2 \cdot 10^{-3}$	2.92	[4]	[8]
$Cu(CSN_2H_4)_3^{2+}$	Room	0 01—0.03	Potent.	—	—	$1.5 \cdot 10^{-13}$	12.82	[1]	—
$Cu(CSN_2H_4)_4^{2+}$	25	0.1	Polarog.	—	—	$4.1 \cdot 10^{-16}$	15.39	[5]	—
$Hg(CSN_2H_4)_2^{2+}$	25	1.0	"	—	—	$1.2 \cdot 10^{-22}$	21.9	[6]	—
$Hg(CSN_2H_4)_3^{2+}$	25	1.0	"	2.10^{-3}	2.7	$3.2 \cdot 10^{-25}$	24.6	[6]	—
$Hg(CSN_2H_4)_4^{2+}$	Room	0.01	Potent.	—	—	$1.1 \cdot 10^{-28}$	27.96	[1]	—
$Hg(CSN_2H_4)_4^{2+}$	25	1.0	Polarog.	2.10^{-2}	1.7	$3\ 2 \cdot 10^{-27}$	26.3	[6]	—
$Pb(CSN_2H_4)_3^{2+}$	25	0.1	"	—	—	$1.7 \cdot 10^{-2}$	1.77	[4]	—

REFERENCES

1. A.T.PILIPENKO and T.S.LISETSKAYA, Ukrain.khim. zhur., 19, 81 (1953).

2. O.S.FEDOROVA, Zh.obshch.khim., 24, 62 (1954).

3. C.L.RULFS, E.R.PRZYBYLOWICZ and C.E.SKINNER, Anal.Chem., 26, 408 (1954).

4. O.S.FEDOROVA, Collected Articles on General Chemistry (Sbornik statei po obshchei khimii), I, 206, Izd.Akad.Nauk SSSR (1953).

5. E.I.ONSTOTT and H.A.LAITINEN, J.Am. Chem.Soc., 72, 4724 (1950).

6. V.F.TOROPOVA, Zh.neorg.khim., 1, 930 (1956).

7. F. G. PAWELKA, Z. Elektrochem., 30, 180 (1924).

8. W.S.FYFE, J.Chem.Soc., 1032 (1955).

Trimetaphosphate Complexes

Complex ion	Temp. °C	Ionic strength	Method	k	pk	K	pK	References
$BaP_3O_9^-$	25	0	El.cond.	$5\ 6\cdot10^{-4}$	3.25	$5.6\cdot10^{-4}$	3.25	[1]
$CaP_3O_9^-$	25	0	Solub. / El.cond	$3\ 4\cdot10^{-4}$	3.47	$3.4\cdot10^{-4}$	3.47	[1,2]
LaP_3O_9	25	0	El.cond.	$2.0\cdot10^{-6}$	5 70	$2.0\cdot10^{-6}$	5.70	[3]
$MgP_3O_9^-$	25	0	.	$3\ 9\cdot10^{-4}$	3.31	$3.9\cdot10^{-4}$	3.31	[1,2]
$MnP_3O_9^-$	25	0	.	$2.8\cdot10^{-4}$	3.56	$2.8\cdot10^{-4}$	3.56	[1]
$NaP_3O_9^{2-}$	25	0	Solub.	$6.8\cdot10^{-2}$	1.17	$6.8\cdot10^{-2}$	1.17	[2]
$NiP_3O_9^-$	25	0	El.cond.	$6\ 0\cdot10^{-4}$	3.22	$6.0\cdot10^{-4}$	3.22	[1]
$SrP_3O_9^-$	25	0	Solub. / El.cond.	$3\ 5\cdot10^{-4}$	3.35	$3.5\cdot10^{-4}$	3.35	[4]

REFERENCES

1. W. H. JONES, C. B. MONK and C. W. DAVIES,
   J. Chem. Soc., 2693 (1949).
2. C. W. DAVIES and C. B. MONK, ibid., 413
   (1949).
3. C. B. MONK, ibid., 1317 (1952).
4. C. B. MONK, ibid., 1314 (1952).

II. COMPLEXES WITH ORGANIC LIGANDS

1. Complexes with Amines

Complexes with Diethylenetriamine $H_2N(CH_2)_2NH(CH_2)_2NH_2$ (Deta)

Complex ion	Temp. °C	Ionic strength	Method	k	pk	K	pK	References principal	References supplementary
AgDeta⁺	20	0.1	pH-potent	$8.0 \cdot 10^{-7}$	6.1	$8.0 \cdot 10^{-7}$	6.1	[1]	—
CdDeta²⁺	20	0.1	"	$3.54 \cdot 10^{-9}$	8.45	$3.54 \cdot 10^{-9}$	8.45	[1]	—
Cd(Deta)₂²⁺	20	0.1	"	$4.0 \cdot 10^{-6}$	5.40	$1.40 \cdot 10^{-14}$	13.85	[1]	[3]
Co Deta²⁺	20	0.1	"	$8.0 \cdot 10^{-9}$	8.1	$8.0 \cdot 10^{-9}$	8.1	[1]	[2]
Co(Deta)₂²⁺	20	0.1	"	$1.0 \cdot 10^{-6}$	6.0	$8.0 \cdot 10^{-15}$	14.1	[1]	[2]
Cu Deta²⁺	20	0.1	"	$1.0 \cdot 10^{-16}$	16.0	$1.0 \cdot 10^{-16}$	16.0	[1]	—
Cu(Deta)₂²⁺	30	1.5	"	$5.0 \cdot 10^{-6}$	5.3	$5.0 \cdot 10^{-22}$	21.3	[1]	—
Fe Deta²⁺	30	1.5	"	$5.9 \cdot 10^{-7}$	6.23	$5.9 \cdot 10^{-7}$	6.23	[2]	—
Fe (Deta)₂²⁺	30	1.5	"	$7.4 \cdot 10^{-5}$	4.13	$4.4 \cdot 10^{-11}$	10.36	[1]	—
HgDeta²⁺	20	0.5	"	$1.6 \cdot 10^{-22}$	21.8	$1.6 \cdot 10^{-22}$	21.8	[1]	—
Hg (Deta)₂²⁺	20	0.5	"	$1.0 \cdot 10^{-7}$	7.0	$1.6 \cdot 10^{-29}$	28.8	[2]	—
Mn Deta²⁺	30	1.5	"	$1.02 \cdot 10^{-4}$	4.0	$1.02 \cdot 10^{-4}$	3.99	[2]	—
Mn (Deta)₂²⁺	30	1.5	"	$1.48 \cdot 10^{-3}$	2.83	$1.51 \cdot 10^{-7}$	6.82	[2]	—
Ni Deta²⁺	30	0.1	"	$2.0 \cdot 10^{-11}$	10.7	$2.0 \cdot 10^{-11}$	10.7	[1]	—
Ni (Deta)₂²⁺	20	0.1	"	$5.6 \cdot 10^{-9}$	8.25	$1.12 \cdot 10^{-19}$	18.95	[1]	[2]
Zn Deta²⁺	20	0.1	"	$1.26 \cdot 10^{-9}$	8.9	$1.26 \cdot 10^{-9}$	8.9	[1]	[2]
Zn (Deta)₂²⁺	20	0.1	"	$3.1G \cdot 10^{-6}$	5.5	$4.0 \cdot 10^{-15}$	14.4	[1]	—

REFERENCES

1. J. E. PRUE and G. SCHWARZENBACH, Helv. chim. Acta, 33, 985 (1950).
2. H. B. JONASSEN, et al., J. Phys. Chem., 56, 16 (1952).
3. B. E. DOUGLAS, H. A. LAITINEN and J. C. BAILAR, J. Am. Chem. Soc., 72, 2484 (1950).

Complexes with Dipyridyl [structure] Dyp

Complex ion	Temp. °C	Ionic strength	Method	k	pk	K	pK	References principal	References supplementary
Cd(Dyp)$_3^{2+}$	25	0.1	Polarog.	—	—	$3.4 \cdot 10^{-11}$	10.47	[1]	[1]
Cu(Dyp)$_{3/2}$	25	0.1	"	—	—	$6.3 \cdot 10^{-15}$	14.2	[2]	—
Cu(Dyp)$_3^{2+}$	25	0.1	"	—	—	$1.4 \cdot 10^{-18}$	17.85	[2]	—
Fe Dyp^{2+}	25	0.025	Kin.	$6.1 \cdot 10^{-5}$	4.21	$6.1 \cdot 10^{-5}$	4.21	[3]	[4,6]
Fe(Dyp)$_2^{2+}$	25	0.025	"	$1.0 \cdot 10^{-5}$	5.0	$6.1 \cdot 10^{-9}$	9.21	[3]	—
Fe(Dyp)$_3^{2+}$	25	0.33	Spectr.	—	—	$2.6 \cdot 10^{-18}$	17.58	[4]	[7]
Mg Dyp^{2+}	25	0.5	"	$3.16 \cdot 10^{-1}$	0.5	$3.16 \cdot 10^{-1}$	0.5	[5]	—
Mn Dyp^{2+}	27	0.5	"	$3.16 \cdot 10^{-3}$	2.5	$3.16 \cdot 10^{-3}$	2.5	[5]	—
Pb Dyp^{2+}	27	0.5	"	$1.0 \cdot 10^{-3}$	3.0	$1.0 \cdot 10^{-3}$	3.0	[5]	—

REFERENCES

1. B. E. DOUGLAS, H. A. LAITTINEN and J. C. BAILAR, Jr., J. Am. Chem. Soc., 72, 2484 (1950).
2. E. I. ONSTOTT and H. A. LAITTINEN, ibid., 72, 4724 (1950).
3. J. H. BAZENDALL and P. GEORGE, Nature, 163, 725 (1949); cited in Chem. Abs., 43, 6056 (1949).
4. P. KRUMHOLZ, Nature, 163, 724 (1949); cited in Chem. Abs., 43, 6056 (1949).
5. K. SONE, P. KRUMHOLZ and H. STAMMREICH, J. Am. Chem. Soc., 77, 777 (1955).
6. F. P. DWYER and R. S. NYHOLM, Proc. Roy. Soc. N. S. Wales, 80, 28 (1946); cited in Chem. Abs., 42, 19 (1948).
7. J. H. BAXENDALL and P. GEORGE, Nature, 162, 777 (1948); cited in Chem. Abs., 43, 3695 (1949).

Complexes with Ethylenediamine $H_2NCH_2CH_2NH_2$ (En)

Complex ion	Temp. °C	Ionic strength	Method	k	pk	K	pK	Reference principal	Reference supplementary
AgEn⁺	16	—	Spectr.	$1 \cdot 10^{-5}$	5.0	$1 \cdot 10^{-5}$	5.0	[1]	—
Ag(En)₂⁺	16	—	"	$1.45 \cdot 10^{-3}$	2.84	$1.45 \cdot 10^{-8}$	7.84	[1]	—
CdEn²⁺	30	0.5	pH-potent.	$3.40 \cdot 10^{-6}$	5.47	$3.40 \cdot 10^{-6}$	5.47	[2]	[6]
Cd(En)₂²⁺	30	0.5	"	$2.82 \cdot 10^{-5}$	4.55	$2.60 \cdot 10^{-11}$	10.02	[2]	[6]
Cd(En)₃²⁺	30	0.5	"	$8.50 \cdot 10^{-3}$	2.07	$8.15 \cdot 10^{-13}$	12.09	[2]	[6]
CoEn²⁺	30	1.2	"	$1.29 \cdot 10^{-6}$	5.89	$1.29 \cdot 10^{-6}$	5.89	[3]	—
Co(En)₂²⁺	30	1.2	"	$1.48 \cdot 10^{-5}$	4.88	$1.91 \cdot 10^{-11}$	10.72	[3]	—
Co(En)₃²⁺	30	1.2	"	$7.95 \cdot 10^{-4}$	3.10	$1.52 \cdot 10^{-14}$	13.82	[3]	—
Co(En)₃³⁺	30	1.0	"	—	—	$2.04 \cdot 10^{-49}$	48.69	[3]	—
CuEn²⁺	30	0.5	"	$2.82 \cdot 10^{-11}$	10.55	$2.82 \cdot 10^{-11}$	10.55	[2]	[4,7]
Cu(En)₂²⁺	30	0.5	"	$8.92 \cdot 10^{-10}$	9.05	$2.52 \cdot 10^{-20}$	19.60	[2]	[4,5,7]
FeEn²⁺	30	1.2	"	$5.25 \cdot 10^{-5}$	4.28	$5.25 \cdot 10^{-5}$	4.28	[3]	—
Fe(En)₂²⁺	30	1.2	"	$5.62 \cdot 10^{-4}$	3.25	$2.95 \cdot 10^{-8}$	7.53	[3]	—
Fe(En)₃²⁺	30	1.2	"	$1.02 \cdot 10^{-2}$	1.99	$3.01 \cdot 10^{-10}$	9.52	[3]	—

Complexes with Ethylenediamine $H_2NCH_2CH_2NH_2$ (En) (contd.)

Complex ion	Temp °C	Ionic strength	Method	k	p^h	K	pK	Reference principal	supplementary
$MnEn^{2+}$	30	1.2	pH-potent.	$1.86 \cdot 10^{-3}$	2.73	$1.86 \cdot 10^{-3}$	2.73	[3]	—
$Mn(En)_2^{2+}$	30	1.2	.	$8.7 \cdot 10^{-3}$	2.06	$1.62 \cdot 10^{-5}$	4.79	[3]	—
$Mn(En)_3^{2+}$	30	1.2	.	$1.32 \cdot 10^{-1}$	0.88	$2.14 \cdot 10^{-6}$	5.67	[3]	—
$NiEn^{2+}$	30	1.2	.	$2.18 \cdot 10^{-8}$	7.66	$2.18 \cdot 10^{-8}$	7.66	[3]	[2,4]
$Ni(En)_2^{2+}$	30	1.2	.	$3.98 \cdot 10^{-7}$	6.40	$8.70 \cdot 10^{-16}$	14.06	[3]	[2,4]
$Ni(En)_3^{2+}$	30	1.2	.	$2.95 \cdot 10^{-5}$	4.53	$2.57 \cdot 10^{-19}$	18.59	[3]	[2,4]
$ZnEn^{2+}$	30	1.2	.	$1.95 \cdot 10^{-6}$	5.71	$1.95 \cdot 10^{-6}$	5.71	[2]	[7]
$Zn(En)_2^{2+}$	30	1.0	.	$2.19 \cdot 10^{-5}$	4.66	$4.26 \cdot 10^{-11}$	10.37	[2]	[7]
$Zn(En)_3^{2+}$	30	1.0	.	$1.91 \cdot 10^{-2}$	1.72	$8.12 \cdot 10^{-13}$	12.08	[2]	—

REFERENCES

1. P. JOB, Ann. chim., [10], 9, 113; cited in Zbl., I, 2572 (1928).
2. G. A. CARLSON, J. P. McREYNOLDS and F. H. VERHOEK, J. Am. Chem. Soc., 67, 1334 (1945).
3. J. BJERRUM, Metal Ammine Formation in Aqueous Solutions, Copenhagen (1941); cited in Chem. Abs., 6527 (1941).
4. F. BASOLO and R. K. MURMAN, J. Am. Chem. Soc., 74, 5243 (1952).
5. E. I. ONSTOTT and H. A. LAITINEN, ibid., 72, 4724 (1950).
6. C. G. SPIKE and R. W. PARRY, ibid., 75, 2726 (1953).
7. C. G. SPIKE and R. W. PARRY, ibid., 75, 3770 (1953).

Complexes with Imidazole $\begin{array}{c} HC - CH \\ | \quad \quad \backslash NH \\ N = CH \end{array}$ Im

Complex ion	Temp. °C	Ionic strength	Method	k	pk	K	pK	References principal	References supplementary
$CdIm^{3+}$	25	0.15	pH-potent	$1.58 \cdot 10^{-3}$	2.80	$1.58 \cdot 10^{-3}$	2.80	[1]	[1]
$Cd(Im)_2^{2+}$	25	0.15	.	$8.0 \cdot 10^{-3}$	2.10	$1.26 \cdot 10^{-5}$	4.90	[1]	[1]
$Cd(Im)_3^{2+}$	25	0.15	.	$2.8 \cdot 10^{-2}$	1.55	$3.5 \cdot 10^{-7}$	6.45	[1]	[1]
$Cd(Im)_4^{2+}$	25	0.15	.	$7.4 \cdot 10^{-2}$	1.13	$2.6 \cdot 10^{-8}$	7.58	[1]	[1]
$CuIm^{2+}$	22.5	0.16	.	$4.4 \cdot 10^{-5}$	4.36	$4.4 \cdot 10^{-5}$	4.36	[2]	[2]
$Cu(Im)_2^{2+}$	22.5	0.16	.	$2.7 \cdot 10^{-4}$	3.57	$1.17 \cdot 10^{-8}$	7.93	[2]	[2]
$Cu(Im)_3^{2+}$	22.5	0.16	.	$1.4 \cdot 10^{-3}$	2.85	$1.65 \cdot 10^{-11}$	10.78	[2]	[2]
$Cu(Im)_4^{2+}$	22.5	0.16	.	$8.7 \cdot 10^{-3}$	2.06	$1.45 \cdot 10^{-13}$	12.84	[2]	[2]
$NiIm^{2+}$	25	0.15	.	$5.4 \cdot 10^{-4}$	3.27	$5.4 \cdot 10^{-4}$	3.27	[2]	—
$Ni(Im)_2^{2+}$	25	0.15	.	$2.1 \cdot 10^{-3}$	2.68	$1.12 \cdot 10^{-6}$	5.95	[3]	—
$Ni(Im)_3^{2+}$	25	0.15	.	$7.1 \cdot 10^{-3}$	2.15	$8.0 \cdot 10^{-9}$	3.10	[3]	—
$Ni(Im)_4^{2+}$	25	0.15	.	$2.2 \cdot 10^{-2}$	1.65	$1.8 \cdot 10^{-10}$	9.75	[3]	—
$Ni(Im)_5^{2+}$	25	0.15	.	$7.6 \cdot 10^{-2}$	1.12	$1.35 \cdot 10^{-11}$	10.87	[3]	—
$Ni(Im)_6^{2+}$	25	0.15	.	$3.0 \cdot 10^{-1}$	0.52	$4.1 \cdot 10^{-12}$	11.39	[3]	—
$ZnIm^{2+}$	24	0.16	.	$2.6 \cdot 10^{-3}$	2.58	$2.6 \cdot 10^{-3}$	2.58	[2]	[2]
$Zn(Im)_2^{2+}$	24	0.16	.	$4.26 \cdot 10^{-8}$	2.37	$1.12 \cdot 10^{-5}$	4.95	[2]	[2]
$Zn(Im)_3^{2+}$	24	0.16	.	$5.9 \cdot 10^{-3}$	2.23	$6.6 \cdot 10^{-8}$	7.18	[2]	[2]
$Zn(Im)_4^{2+}$	24	0.16	.	$9.5 \cdot 10^{-3}$	2.02	$6.3 \cdot 10^{-10}$	9.20	[2]	[2]

REFERENCES

1. C. TANFORD and M. L. WAGNER, J. Am. Chem. Soc., 75, 434 (1953).
2. J. T. EDSALL, et al., ibid., 76, 3054 (1954).
3. N. C. LI, et al., 77, 859 (1955).

Complexes with Phenanthroline ⟨N⟩⟨N⟩ (Ph)

Complex ion	Temp. °C	Ionic strength	Method	k	p^k	K	p^K	References principal	References supplementary
CaPh²⁺	27	0.5	Spectr.	$3.2 \cdot 10^{-1}$	0.5	$3.2 \cdot 10^{-1}$	0.5	[1]	—
Cd(Ph)₂²⁺	25	0.03*	Polarog.	—	—	$7.0 \cdot 10^{-14}$	13.16	[2]	—
Cd(Ph)₃²⁺	25	0.02	"	—	—	$6.4 \cdot 10^{-16}$	15.20	[2]	—
FePh³⁺	25	0.7	Spectr.	$1.3 \cdot 10^{-6}$	5.89	$1.3 \cdot 10^{-6}$	5.89	[3]	—
Fe(Ph)₃²⁺	25	0	Potent.	—	—	$5.0 \cdot 10^{-22}$	21.3	[4]	[6,7]
MgPh²⁺	27	0.5	Spectr.	$3.2 \cdot 10^{-2}$	1.5	$3.2 \cdot 10^{-2}$	1.5	[1]	—
ZnPh²⁺	25	0.1	Distrib.	$3.7 \cdot 10^{-7}$	6.43	$3.7 \cdot 10^{-7}$	6.43	[5]	—
Zn(Ph)₂²⁺	25	0.1	"	$1.9 \cdot 10^{-6}$	5.72	$7 \cdot 10^{-18}$	12.15	[5]	—
Zn(Ph)₃²⁺	25	0.1	"	$1.6 \cdot 10^{-5}$	4.8	$1 \cdot 10^{-17}$	17.0	[5]	—

* 28.5% ethanol.

REFERENCES

1. K. SONE, P. KRUMHOLZ and H. STAMMERICH, J. Am. Chem. Soc., 77, 777 (1955).
2. B. E. DOUGLAS, H. A. LAITINEN and J. C. BAILAR Jr., ibid., 72, 2484 (1950).
3. I. M. KOLTHOFF, D. L. LEUSSING and T. S. LEE, ibid., 72, 2173 (1950).
4. T. S. LEE, I. M. KOLTHOFF and D. L. LEUSSING, ibid., 70, 2348 (1948).
5. I. M. KOLTHOFF, D. L. LEUSSING and T. S. LEE, ibid., 73, 390 (1951).
6. C. M. COOK and F. A. LONG, ibid., 73, 4119 (1951).
7. F. P. DWYER and R. S. NYHOLM, J. Proc. Roy. Soc. N. S. Wales, 80, 28 (1946); cited in Chem. Abs., 43, 19 (1948).

Complexes with Propylenediamine $H_2NCHCH_2NH_2$ (Pn)
$$\underset{CH_3}{|}$$

Complex ion	Temp. °C	Ionic strength	Method	k	pk	K	pK	References principal	References supplementary
CdPn²⁺	30	0.5	pH-potent.	$3.80 \cdot 10^{-6}$	4.52	$3.80 \cdot 10^{-6}$	5.42	[1]	—
Cd(Pn)₂²⁺	30	0.5	"	$2.82 \cdot 10^{-5}$	4.55	$1.07 \cdot 10^{-10}$	9.97	[1]	—
Cd(Pn)₃²⁺	30	0.5	"	$7.08 \cdot 10^{-3}$	2.15	$7.59 \cdot 10^{-13}$	12.12	[1]	[2]
CuPn²⁺	30	0.5	"	$2.63 \cdot 10^{-11}$	10.58	$2.63 \cdot 10^{-11}$	10.58	[1]	—
Cu(Pn)₂²⁺	30	0.5	"	$8.31 \cdot 10^{-10}$	9.08	$2.19 \cdot 10^{-20}$	19.66	[1]	—
NiPn²⁺	30	0.5	"	$3.90 \cdot 10^{-8}$	7.41	$3.90 \cdot 10^{-8}$	7.41	[1]	—
Ni(Pn)₂²⁺	30	0.5	"	$5.00 \cdot 10^{-7}$	6.30	$1.95 \cdot 10^{-14}$	13.71	[1]	—
Ni(Pn)₃²⁺	30	0.5	"	$5.13 \cdot 10^{-5}$	4.29	$1.0 \cdot 10^{-18}$	18.00	[1]	—
ZnPn²⁺	30	0.5	"	$1.29 \cdot 10^{-6}$	5.89	$1.29 \cdot 10^{-6}$	5.89	[1]	—
Zn(Pn)₂²⁺	30	0.5	"	$1.05 \cdot 10^{-5}$	4.98	$1.35 \cdot 10^{-11}$	10.87	[1]	—
Zn(Pn)₃²⁺	30	0.5	"	$2.00 \cdot 10^{-2}$	1.70	$2.70 \cdot 10^{-13}$	12.57	[1]	—

REFERENCES

1. G. A. CARLSON, J. P. McREYNOLDS and F. H. VERHOEK, J. Am. Chem. Soc., 67, 1334 (1945).

2. B. E. DOUGLAS, H. A. LAITINEN and J. C. BAILAR, Jr., ibid., 72, 2484 (1950).

Complexes with Pyridine ⟨Pyr⟩

Complex ion	Temp. °C	Ionic strength	Method	k	pk	K	pK	References principal	References supplementary
AgPyr⁺	25	—	Solub.	$1.0 \cdot 10^{-2}$	2.0	$1.0 \cdot 10^{-2}$	2.0	[1]	[4]
Ag(Pyr)₂⁺	25	—	"	$7.8 \cdot 10^{-3}$	2.11	$7.8 \cdot 10^{-3}$	4.11	[1]	[4]
Cd(Pyr)₂²⁺	25	0.1	Polarog.	—	—	$7.2 \cdot 10^{-3}$	2.14	[2]	—
Cd Pyr₄²⁺	25	0.1	"	—	—	$3.2 \cdot 10^{-3}$	2.49	[2]	—
Cu(Pyr)₂⁺	25	0.01	"	$6.75 \cdot 10^{-2}$	1.17	$4.6 \cdot 10^{-1}$	3.34	[3]	—
Cu(Pyr)₃	25	0.01	"	$1.17 \cdot 10^{-1}$	0.93	$3.1 \cdot 10^{-5}$	4.51	[3]	—
Cu(Pyr)₄⁺	25	0.01	"	—	—	$3.6 \cdot 10^{-6}$	5.44	[3]	—
Cu(Pyr)₆	25	0.01	"	—	—	$1.3 \cdot 10^{-7}$	6.89	[3]	—
CuPyr²⁺	25	0.5	pH-potent.	$3.02 \cdot 10^{-3}$	2.52	$3.02 \cdot 10^{-3}$	2.52	[4]	—
Cu(Pyr)₂²⁺	25	0.5	"	$1.38 \cdot 10^{-1}$	1.86	$4.16 \cdot 10^{-5}$	4.38	[4]	—
Cu(Pyr)₃²⁻	25	0.5	"	$4.90 \cdot 10^{-2}$	1.31	$2.04 \cdot 10^{-6}$	5.69	[4]	—
Cu(Pyr)₄²⁺	25	0.5	"	0.141	0.85	$2.88 \cdot 10^{-7}$	6.54	[4]	—
Cu(Pyr)₆²⁻	25	0.01	Polarog.	—	—	0.1	1.0	[3]	—

REFERENCES

1. W. C. VOSBURGH and S. A. COGSWELL, J. Am. Chem. Soc., 65, 2412 (1943).
2. E. DOUGLAS, H. A. LAITINEN and J. C. BAILAR, ibid., 72, 2484 (1950).
3. I. A. KORSHUNOV and N. I. MALYUGINA, Zh. obshch. khim., 20, 402 (1950).
4. R. J. BRUEHLMAN and F. H. VERHOEK, J. Am. Chem. Soc., 70, 1401 (1948).

Complexes with 1, 2, 3-Triaminopropane $C_3H_5(NH_2)_3$ (Ptn)

Complex ion	Temp. °C	Ionic strength	Method	k	pk	K	pK
AgPtn⁺	20	0.1	pH-potent.	$2.24 \cdot 10^{-6}$	5.65	$2.24 \cdot 10^{-6}$	5.65
CdPtn²⁺	20	0.1	"	$3.55 \cdot 10^{-7}$	6.45	$3.55 \cdot 10^{-7}$	6.45
CoPtn³⁺	20	0.1	"	$1.6 \cdot 10^{-7}$	6.8	$1.6 \cdot 10^{-7}$	6.8
CuPtn²⁺	20	0.1	"	$8.0 \cdot 10^{-13}$	11.1	$8.0 \cdot 10^{-12}$	11.1
Cu(Ptn)₂²⁺	20	0.1	"	$1.0 \cdot 10^{-9}$	9.0	$8.0 \cdot 10^{-21}$	20.1
HgPtn²⁺	20	0.5	"	$2.5 \cdot 10^{-20}$	19.6	$2.5 \cdot 10^{-20}$	19.6
NiPtn³⁺	20	0.1	"	$5.0 \cdot 10^{-10}$	9.3	$5.0 \cdot 10^{-10}$	9.3
ZnPtn²⁺	20	0.1	"	$1.78 \cdot 10^{-7}$	6.75	$1.78 \cdot 10^{-7}$	6.75

REFERENCE

.: J. E. PRUE and G. SCHWARZENBACH, Helv. Chim. Acta, 33, 995 (1950).

Complexes with Methylamine CH_3NH_2

Complex ion	Temp. °C	Ionic strength	Method	k	pk	K	pK
Cd(CH₃NH₂)²⁺	25	1.0	pH-potent.	$1.80 \cdot 10^{-3}$	2.745	$1.80 \cdot 10^{-3}$	2.745
Cd(CH₃NH₂)₂²⁺	25	1.0	"	$8.66 \cdot 10^{-3}$	2.063	$1.56 \cdot 10^{-5}$	4.808
Cd(CH₃NH₂)₃²⁺	25	1.0	"	$7.40 \cdot 10^{-2}$	1.131	$1.15 \cdot 10^{-6}$	5.939
Cd(CH₃NH₂)₄²⁺	25	1.0	"	0.245	0.611	$2.82 \cdot 10^{-7}$	6.550

REFERENCE

1. C. G. SPIKE and R. W. PARRY, J. Am. Chem. Soc., 75, 2726 (1953).

Complexes with Triaminotriethylamine N $(CH_2CH_2NH_2)_3$ (Tate)

Complex ion	Temp. °C	Ionic strength	Method	k	pk	K	pK
Ag Tate⁺	20	0.1	pH-potent	$1.58 \cdot 10^{-8}$	7.8	$1.58 \cdot 10^{-8}$	7.8
Cd Tatc³⁺	20	0.1	"	$5.0 \cdot 10^{-13}$	12.3	$5.0 \cdot 10^{-13}$	12.3
Co Tate²⁺	20	0.1	"	$1.58 \cdot 10^{-13}$	12.8	$1.58 \cdot 10^{-13}$	12.8
Cu Tate²⁺	20	0.1	"	$1.58 \cdot 10^{-19}$	18.8	$1.58 \cdot 10^{-19}$	18.8
Fe Tate²⁺	20	0.1	"	$1.58 \cdot 10^{-9}$	8.8	$1.58 \cdot 10^{-9}$	8.8
Hg Tate²⁺	20	0.5	"	$1.58 \cdot 10^{-23}$	22.8	$1.58 \cdot 10^{-23}$	22.8
Mn Tate²⁺	20	0.1	"	$1.58 \cdot 10^{-6}$	5.8	$1.58 \cdot 10^{-6}$	5.8
Ni Tate²⁺	20	0.1	"	$1.58 \cdot 10^{-15}$	14.8	$1.58 \cdot 10^{-15}$	14.8
Zn Tate²⁺	20	0.1	"	$2.24 \cdot 10^{-15}$	14.65	$2.24 \cdot 10^{-15}$	14.65

REFERENCE

1. J. E. PRUE and G. SCHWARZENBACH, Helv. chim. Acta, 33, 963 (1950).

Complexes with Trimethylenediamine $H_2NCH_2CH_2CH_2NH_2$ (Tmen)

Complex ion	Temp. °C	Ionic strength	Method	k	pk	K	pK
CuTmen²⁺	25	1.0	pH-potent	$1.05 \cdot 10^{-10}$	9.98	$1.05 \cdot 10^{-10}$	9.98
Cu(Tmen)₂²⁺	25	1.0	"	$6.45 \cdot 10^{-8}$	7.19	$6.75 \cdot 10^{-18}$	17.17
NiTmen²⁺	25	1.0	"	$4.07 \cdot 10^{-7}$	6.39	$4.07 \cdot 10^{-7}$	6.39
Ni(Tmen)₂²⁺	25	1.0	"	$4.07 \cdot 10^{-5}$	4.39	$1.66 \cdot 10^{-11}$	10.78
Ni(Tmen)₃²⁺	25	1.0	"	$5.90 \cdot 10^{-2}$	1.23	$9.80 \cdot 10^{-13}$	12.01

REFERENCE

1. I. POULSEN and J. BJERRUM, Acta Chem. Scand., 2, 1407 (1955); cited in Ref. zh. khim., 57786 (1956).

Complexes with Triethylenetetramine $NH_2(CH_2)_2NH(CH_2)_2NH(CH_2)_2NH_2$ (Teta)

Complex ion	Temp. °C	Ionic strength	Method	k	pk	K	pK	Reference
Ag Teta⁺	20	0.1	pH-potent.	$2.0 \cdot 10^{-8}$	7.7	$2.0 \cdot 10^{-8}$	7.7	[1]
Cd Teta²⁺	20	0.1		$1.78 \cdot 10^{-11}$	10.75	$1.78 \cdot 10^{-11}$	10.75	[1]
Cd (Teta)₂²⁺	25	0.3	Polarog.	—	—	$1\,2 \cdot 10^{-14}$	13.9	[2]
Co Teta²⁺	20	0.1	pH-potent.	$1.0 \cdot 10^{-11}$	11.0	$1.0 \cdot 10^{-11}$	11.0	[1]
Cu Teta²⁺	20	0.1	•	$4.0 \cdot 10^{-21}$	20.4	$4.0 \cdot 10^{-21}$	20.4	[1]
Fe Teta²⁺	20	0.1	•	$1.6 \cdot 10^{-8}$	7.8	$1.6 \cdot 10^{-8}$	7.8	[1]
Hg Teta²⁺	20	0.5	•	$5.5 \cdot 10^{-26}$	25.26	$5.5 \cdot 10^{-26}$	25.26	[1]
Mn Teta²⁺	20	0.1	•	$1.26 \cdot 10^{-5}$	4.9	$1.26 \cdot 10^{-5}$	4.9	[1]
Ni Teta²⁺	20	0.1	•	$1.0 \cdot 10^{-14}$	14.0	$1.0 \cdot 10^{-14}$	14.0	[1]
Zn Teta²⁺	20	0.1	•	$8.0 \cdot 10^{-13}$	12.1	$8.0 \cdot 10^{-13}$	12.1	[1]

REFERENCES

1. G. SCHWARZENBACH, Helv. chim. Acta, 33, 974 (1950).

2. B. E. DOUGLAS, H. A. LAITINEN and J. C. BAILAR, Jr., J. Am. Chem. Soc., 72, 2484 (1950).

2. Complexes with Organic Acid Anions

Acetate Complexes

Complex ion	Temp. °C	Ionic strength	Method	k	pk	K	pK	References principal	References supplementary
AgCH$_3$COO	25	2.95	Potent.	0.183	0.74	0.183	0.74	[1]	—
Ag(CH$_3$COO)$_2^-$	25	2.95	"	1.26	0.10	0.230	0.64	[1]	—
BaCH$_3$COO$^+$	25	0	Solub.	0.39	0.41	0.39	0.41	[2]	[16]
"	—	0.20	Potent.	0.42	0.38	0.42	0.38	[3]	—
CaCH$_3$COO$^+$	25	0	Solub.	0.17	0.77	0.17	0.77	[2]	[16,17]
"	25	0.20	Potent.	0.29	0.53	0.29	0.53	[3]	—
CdCH$_3$COO$^+$	—	—	"	$1.7 \cdot 10^{-2}$	1.77	$1.7 \cdot 10^{-2}$	1.77	[4]	—
Cd(CH$_3$COO)$_2$	—	—	"	$1.0 \cdot 10^{-1}$	1.0	$1.7 \cdot 10^{-3}$	2.77	[4]	—
CeCH$_3$COO^{2+}	20	1.00	Ion exch.	$2.1 \cdot 10^{-2}$	1.68	$2.7 \cdot 10^{-2}$	1.68	[5]	—
Ce(CH$_3$COO)$_2^+$	20	1.00	"	0.107	0.97	$2.2 \cdot 10^{-3}$	2.65	[5]	—
Ce(CH$_3$COO)$_3$	20	1.0	"	0.21	0.68	$5.9 \cdot 10^{-4}$	3.23	[5]	—
CuCH$_3$COO$^+$	25	0	Solub.	$5.7 \cdot 10^{-3}$	2.24	$5.7 \cdot 10^{-3}$	2.24	[6]	[7]
Cu(CH$_3$COO)$_2$	—	—	Potent.	0.125	0.90	$5.0 \cdot 10^{-4}$	3.30	[7]	—
Hg(CH$_3$COO)$_2$	—	—	Solub.	—	—	$3.75 \cdot 10^{-9}$	8.43	[8]	—
InCH$_3$COO^{2+}	—	—	Potent.	$3.12 \cdot 10^{-4}$	3.51	$3.12 \cdot 10^{-4}$	3.51	[9]	—

Acetate Complexes (contd.)

Complex ion	Temp. °C	Ionic strength	Method	k	pk	K	pK	References principal	supplementary
In(H₃COO)₂⁺	—	—	Potent.	$3.6 \cdot 10^{-3}$	2.44	$1.11 \cdot 10^{-6}$	5.95	[9]	—
In(CH₃COO)₃	—	—	"	$1.12 \cdot 10^{-2}$	1.95	$1.25 \cdot 10^{-8}$	7.90	[9]	—
In(CH₃COO)₄⁻	—	—	"	$6.6 \cdot 10^{-2}$	1.18	$8.3 \cdot 10^{-10}$	9.08	[9]	—
In(CH₃COO)₅²⁻	—	—	"	0.71	0 15	$5.9 \cdot 10^{-10}$	9.23	[9]	—
In(CH₃COO)₆³⁻	—	—	"	$8.5 \cdot 10^{-2}$	1.07	$5.0 \cdot 10^{-11}$	18.30	[9]	—
MgCH₃COO⁺	25	0	"	0.165	0.82	0.165	0.82	[2]	[16]
MnCH₃COO⁺	25	0	Calc.	$6 \cdot 10^{-2}$	1.2	$6 \cdot 10^{-2}$	1.2	[10]	—
NiCH₃COO⁺	25	0	"	$1.6 \cdot 10^{-2}$	1.8	$1.6 \cdot 10^{-2}$	1.8	[10]	[18]
"	20	1.0	Potent.	0.22	0.67	0.22	0.67	[11]	—
Ni(CH₃COO)₂	20	1.0	"	0.26	0.59	$5.5 \cdot 10^{-2}$	1.26	[11]	—
PbCH₃COO⁺	25	1.0	Solub.	$9 \cdot 10^{-3}$	2.05	$9.0 \cdot 10^{-3}$	2.05	[12]	[19]
"	20	2.0	Polarog.	$1.6 \cdot 10^{-2}$	1.80	$1.6 \cdot 10^{-2}$	1.80	[13,14]	—
Pb(CH₃COO)₂	20	2.0	"	0.58	0.24	$0.92 \cdot 10^{-2}$	2.04	[13,14]	—
Pb(CH₃COO)₃⁻	20	2.0	"	1.35	−0.13	$1.2 \cdot 10^{-2}$	1.91	[13,14]	—
Pb(CH₃COO)₄²⁻	20	2.0	"	3.16	−0.50	$3.8 \cdot 10^{-2}$	1.41	[13,14]	[16,17]
SrCH₃COO⁺	25	0	Solub.	0.36	0.44	0.36	0.44	[2]	[3]
ZnCH₃COO⁺	18	0.1	Potent.	$2 \cdot 10^{-2}$	1.70	$2 \cdot 10^{-3}$	1.70	[15]	

REFERENCES

1. F.H.MacDOUGALL and L.E.TOPOL, J.Phys.Chem., 56, 1090 (1952).
2. C.A.COLMAN-PORTER and C.B.MONK, J.Chem.Soc., 4363 (1952).
3. R.K.CANNAN and A.KIBRICK, J.Am.Chem.Soc., 60, 2314 (1938).
4. S.ADITYA and B.PRASAD, J.Indian Chem.Soc., 30, 255 (1953);
   Chem.Abs., 48, 488 (1954).
5. S.FRONAEUS, Svensk.Kem.Tidsk., 65, 19 (1953); Chem.Abs.,
   47, 8459 (1953).
6. M.LLOYD, V.WYCHERLEY and C.B.MONK, J.Chem.Soc., 1786 (1951).
7. S.S.SORCAR, S.ADITYA and B.PRASAD, J.Indian, Chem.Soc.,
   30, 255 (1953).
8. P.MAHAPATRA, S.ADITYA and B.PRASAD, ibid., 30, 509 (1953).
9. N.SUNDEN, Svensk.Kem.Tidsk., 65, 257 (1953); Ref.zh.
   khim., 13786 (1955).
10. K.B.YATSIMIRSKII, Zh.obshch.khim., 24, 1498 (1954).
11. S.FRONAEUS, Acta Chem.Scand., 6, 1200 (1952); Chem.Abs.,
    47, 5292 (1953).
12. S.M.EDMONDS and N.BRINBAUM, J.Am.Chem.Soc., 62, 2367 (1940).
13. V.F.TOROPOVA and F.M.BATORGSHINA, Zh.anal.khim., 4, 337
    (1949).
14. K.B.YATSIMIRSKII, Collected Articles on General Chemistry
    (Sb. statei po obshchei khimii), I, 193, Izd. Akad.
    Nauk SSSR (1953).
15. E.FERRELL, J.M.RIDGION and H.L.RILEY, J.Chem.Soc., 1121
    (1936).
16. G.H.NANCOLLAS, ibid., 744 (1956).
17. J.SCHUBERT and A.LINDENBAUM, J.Am.Chem.Soc., 74, 3529
    (1952).
18. P.K.JENA, S.A.ADITYA and B.PRASAD, J.Indian, Chem.Soc.,
    30, 735 (1953); Chem.Abs., 48, 9161 (1954).
19. N.K.DAS, S.ADITYA and B.PRASAD, Chem.Abs., 46, 10802
    (1952).

Butyrate Complexes $C_3H_7COO^-$ (Bu$^-$)

Complex ion	Temp. °C	Ionic strength	Method	h	pk	K	pK
BaBu⁺	—	0.2	pH-potent.	$4.90 \cdot 10^{-1}$	0.31	$4.90 \cdot 10^{-1}$	0.31
CaBu⁺	—	0.2	—	$3.10 \cdot 10^{-1}$	0.51	$3.10 \cdot 10^{-1}$	0.51
MgBu⁺	—	0.2	"	$2.95 \cdot 10^{-1}$	0.53	$2.95 \cdot 10^{-1}$	0.53
SrBu⁺	—	0.2	"	$4.35 \cdot 10^{-1}$	0.36	$4.35 \cdot 10^{-1}$	0.36
ZnBu⁺	—	0.2	"	$1.0 \cdot 10^{-1}$	1.00	$1.0 \cdot 10^{-1}$	1.00

REFERENCE

1. R.K.CANNAN and A.KIBRICK, J.Am.Chem.Soc., 63, 2314 (1951).

Citrate Complexes $C_3H_4OH(COO)_3^{3-}$ (Cit^{3-})

Complex ion	Temp °C	Ionic strength	Method	k	pk	K	pK	References principal	References supplementary
BaCit⁻	25	0.16	Ion exch.	$5\cdot10^{-3}$	3	$5\cdot10^{-3}$	2.3	[1]	[1,3]
BeCit⁻	34	0.15	"	$3\cdot10^{-5}$	4.52	$3\cdot10^{-5}$	4.52	[2]	—
BeHCit	34	0.15	"	$6\cdot10^{-3}$	2.22	$6\cdot10^{-3}$	2.22	[2]	—
BeH₂Cit⁺	34	0.15	"	$4\cdot10^{-2}$	1.40	$4\cdot10^{-2}$	1.40	[2]	—
CaCit⁻	25	0.15	Potent.	$6.75\cdot10^{-4}$	3.17	$6.75\cdot10^{-4}$	3.17	[3]	[4,12]
CaHCit	25	0	Ion exch.	$8.1\cdot10^{-4}$	3.09	$8.1\cdot10^{-4}$	3.09	[4]	—
CaH₂Cit⁺	25	0	"	$8\cdot10^{-2}$	1.90	$8\cdot10^{-2}$	1.90	[4]	—
CdCit⁻	25	0.1	Polarog.	$6\cdot10^{-5}$	4.22	$6\cdot10^{-5}$	4.22	[5]	—
CdCitC₄H²⁻	25	0.1	"	$9\cdot10^{-6}$	5.05	$5\cdot10^{-10}$	9.30	[5]	—
Ce(H₂Cit)₃	25	0.5	Ion exch.	—	—	$6.3\cdot10^{-4}$	3.2	[6]	—
CuCir⁻	25	>0.5	Polarog.	$6.2\cdot10^{-15}$	14.21	$6.2\cdot10^{-15}$	14.21	[7]	—
Cu(OH)₂(Cit)₂⁶⁻	25	>0.5	"	—	—	$5\cdot10^{-20}$	19.30	[7]	—
Ca(H₂Cit)(HCit)⁻	room	—	Spectr.	—	—	$1\cdot10^{-4}$	4.0	[8]	—
Cu(OH)(Cit)⁴⁻	·	—	Solub.	—	—	$4.5\cdot10^{-17}$	16.35	[8]	—

Citrate Complexes (contd.)

Complex ion	Temp. °C	Ionic strength	Method	k	p_k	K	pK	References principally	supplementary
$Cu(OH)_2(Cit)_2^{6-}$	.	—	Solub.	$3.8\cdot10^{-3}$	2.42	$1.7\cdot10^{-19}$	18.77	[8]	—
$MgCit^-$	25	0.16	Biol.	$6.3\cdot10^{-4}$	3.2	$6.3\cdot10^{-4}$	3.2	[9]	—
$PbCit^-$	25	0.16	Potent.	$1.8\cdot10^{-6}$	5.74	$1.8\cdot10^{-6}$	5.74	[10]	—
$Pr(H_2Cit)_3$	25	0.5	Ion exch.	—	—	$4.4\cdot10^{-4}$	3.4	[6]	—
$RaCit^-$	25	0.16	Ion exch.	$1\cdot10^{-2}$	2.0	$1\cdot10^{-2}$	2.0	[11]	—
$SrCit^-$	25	0.15	Potent.	$1.2\cdot10^{-3}$	2.92	$1.2\cdot10^{-3}$	2.92	[3]	[1,12,13]
$Y(H_2Cit)_3$	25	0.5	Ion exch.	—	—	$2.3\cdot10^{-4}$	3.6	[6]	—

REFERENCES

1. J.SCHUBERT and J.W.RICHTER, J.Am.Chem.Soc., 70, 4259 (1948).
2. I.FELDMAN, et al., ibid., 77, 878 (1955).
3. N.R.JOSEPH, J.Biol.Chem., 164, 529 (1946); cited in Chem.Abs., 40, 6321 (1946).
4. C.W.DAVIES and B.E.HOYLE, J.Chem.Soc., 1038 (1955).
5. L.MEITES, J.Am.Chem.Soc., 73, 3727 (1951).
6. E.R.TOMPKINS and S.W.MAYER, ibid., 69, 2859 (1947).
7. L.MEITES, ibid., 72, 180 (1950).
8. O.D.TALALAYEVA and A.S.TIKHONOV, Zh. obshch.khim., 23, 2067 (1953).

9. A.B.HASTINGS, et al., J.Biol.Chem., 107, 351 (1934); cited in A.E.Martell, M.Calvin, Chemistry of the Metal Chelate Compounds, New York (1953).
10. S.S.KETY, J.Biol.Chem., 142, 181 (1942).
11. A.E.MARTELL and M.CALVIN, Chemistry of the Metal Chelate Compounds, New York (1953).
12. J.SCHUBERT and A.LINDENBAUM, J.Am.Chem. Soc., 74, 3529 (1952).
13. J.SCHUBERT and J.W.RICHTER, J.Phys. Colloid Chem., 52, 350 (1948); cited in Chem.Abs., 42, 5301 (1948).

Glycerate Complexes $CH_2OHCHOHCOO^-$ (Glac$^-$)

Complex ion	Temp. °C	Ionic strength	Method	k	pk	K	pK
BaGlac$^+$	—	0.2	pH-potent.	$1.6 \cdot 10^{-1}$	0 80	$1.6 \cdot 10^{-1}$	0.80
CaGlac$^+$	—	0.2	"	$6.6 \cdot 10^{-2}$	1.18	$6.6 \cdot 10^{-2}$	1.18
MgGlac$^+$	—	0.2	"	$1.4 \cdot 10^{-1}$	0.86	$1.4 \cdot 10^{-1}$	0.86
SrGlac$^+$	—	0.2	"	$1.3 \cdot 10^{-1}$	0.89	$1.3 \cdot 10^{-1}$	0.89
ZnGlac$^+$	—	0.2	"	$1.6 \cdot 10^{-2}$	1.80	$1.6 \cdot 10^{-2}$	1.80

REFERENCE

1. R.K.CANNAN and A.KIBRICK, J.Am.Chem.Soc., 60, 2314 (1938).

Glycollate Complexes $HOCH_2COO^-$ (Glyk$^-$)

Complex ion	Temp. °C	Ionic strength	Method	k	pk	K	pK	References
BaGlyk$^+$	—	0.2	pH-potent.	$2.18 \cdot 10^{-1}$	0.66	$2.18 \cdot 10^{-1}$	0.66	[1]
CaGlyk$^+$	—	0.2	"	$7.76 \cdot 10^{-2}$	1.11	$7.76 \cdot 10^{-2}$	1.11	[1]
MgGlyk$^+$	—	0.2	"	$1.2 \cdot 10^{-1}$	0.92	$1.2 \cdot 10^{-1}$	0.92	[1]
SrGlyk$^+$	—	0.2	"	$1.58 \cdot 10^{-1}$	0.80	$1.58 \cdot 10^{-1}$	0.80	[1]
UO$_2$Glyk$^+$	—	—	"	$3.8 \cdot 10^{-3}$	2.42	$3.8 \cdot 10^{-3}$	2.42	[2]
UO$_2$(Glyk)$_2$	—	—	"	$2.9 \cdot 10^{-2}$	1.54	$1.1 \cdot 10^{-1}$	3.96	[2]
'$_2$(Glyk)$_3^-$	—	—	"	$5.7 \cdot 10^{-2}$	1.24	$6.3 \cdot 10^{-6}$	5.20	[2]
ZnGlyk$^+$	—	0.2	"	$1.2 \cdot 10^{-2}$	1.92	$1.2 \cdot 10^{-2}$	1.92	[1]

REFERENCES

1. R.K.CANNAN and A.KIBRICK, J.Am.Chem.Soc., 60, 2314 (1938).
2. S.AHRLAND, Acta Chem.Scand., 7, 485 (1953).

Gluconate Complexes $CH_2OH(CHOH)_4COO^-$ Glu^-

Complex ion	Temp. °C	Ionic strength	Method	k	pk	K	pK	References principal	References supplementary
BaGlu⁺	—	0.2	pH-potent.	$1.12 \cdot 10^{-1}$	0.95	$1.12 \cdot 10^{-1}$	0.95	[1]	—
CaGlu⁺	25	0.16	Ion exch.	$6.03 \cdot 10^{-2}$	1.22	$6.03 \cdot 10^{-2}$	1.22	[2]	[1]
MgGlu⁺	—	0.2	pH-potent.	$2.0 \cdot 10^{-1}$	0.70	$2.0 \cdot 10^{-1}$	0.70	[1]	—
SrGlu⁺	25	0.16	Ion exch.	$9.8 \cdot 10^{-2}$	1.01	$9.8 \cdot 10^{-2}$	1.01	[2]	[1]
ZnGlu⁺	—	0.2	pH-potent.	$2.0 \cdot 10^{-2}$	1.70	$2.0 \cdot 10^{-2}$	1.70	[1]	—

REFERENCES

1. R. K. CANNAN and A. KIBRICK, J. Am. Chem. Soc., 60, 2314 (1938).

2. J. SCHUBERT and A. LINDENBAUM, ibid., 74, 3529 (1952).

Complexes with the Kojic Acid Anion * (Coy⁻)

Complex ion	Temp. °C	Ionic strength	Method	k	pk	K	pK
CaCoy⁺	30	—	pH-potent.	$4.0 \cdot 10^{-5}$	4.4	$4.0 \cdot 10^{-5}$	4.4
Ca(Coy)₂	30	—	"	$2.0 \cdot 10^{-3}$	2.7	$8.0 \cdot 10^{-8}$	7.1
CdCoy⁺	30	—	"	$2.5 \cdot 10^{-7}$	6.6	$2.5 \cdot 10^{-7}$	6.6
Cd(Coy)₂	30	—	"	$2.0 \cdot 10^{-5}$	4 7	$5.0 \cdot 10^{-12}$	11.3
CoCoy⁺	30	—	"	$1.6 \cdot 10^{-7}$	6.8	$1.6 \cdot 10^{-7}$	6.8
Co(Coy)₂	30	—	"	$6.3 \cdot 10^{-6}$	5.2	$1.0 \cdot 10^{-12}$	12.0
CuCoy⁺	30	—	"	$5.0 \cdot 10^{-10}$	9.3	$5.0 \cdot 10^{-10}$	9.3
Cu(Coy)₂	30	—	"	$6.3 \cdot 10^{-8}$	7.2	$3.2 \cdot 10^{-17}$	16.5
NiCoy⁺	30	—	"	$8.0 \cdot 10^{-8}$	7.1	$8.0 \cdot 10^{-8}$	7.1
Ni(Coy)₂	30	—	"	$3.16 \cdot 10^{-6}$	5.5	$2.5 \cdot 10^{-13}$	12.6
UO₂Coy⁺	30	—	"	$8.0 \cdot 10^{-11}$	10.1	$8.0 \cdot 10^{-11}$	10.1
UO₂(Coy)₂	30	—	"	$4.0 \cdot 10^{-8}$	7.4	$3.2 \cdot 10^{-18}$	17.5
ZnCoy⁺	30	—	"	$4.0 \cdot 10^{-8}$	7.4	$4.0 \cdot 10^{-8}$	7.4
Zn(Coy)₂	30	—	"	$1.6 \cdot 10^{-6}$	5.8	$6.3 \cdot 10^{-14}$	13.2

* In 50% dioxan-water.

REFERENCES

1. B. E. BRYANT and W. C. FERNELIUS, J. Am. Chem. Soc., 76, 5351 (1954).

Lactate Complexes $CH_3CHOHCOO^-$ (Lac^-)

Complex ion	Temp. °C	Ionic strength	Method	k	pk	K	pK	References principal	supplementary
BaLac⁺	—	0.2	pH-potent.	$2.8 \cdot 10^{-1}$	0.55	$2.8 \cdot 10^{-1}$	0.55	[1]	—
CaLac⁺	25	0.15	"	$1.5 \cdot 10^{-1}$	0.82	$1.5 \cdot 10^{-1}$	0.82	[2]	[1,5,6]
Co(Lac)₂	—	—	Spectr.	—	—	$2.08 \cdot 10^{-2}$	1.68	[3]	—
Cu(Lac)₂	—	—	"	—	—	$2.05 \cdot 10^{-3}$	2.69	[4]	—
MgLac⁺	—	0.2	pH-potent	$1.17 \cdot 10^{-1}$	0.93	$1.17 \cdot 10^{-1}$	C 93	[1]	—
SrLac⁺	—	0.2	"	$2.0 \cdot 10^{-1}$	0.70	$2.0 \cdot 10^{-1}$	0.7c	[1]	[6]
ZnLac⁺	—	0.2	"	$1.38 \cdot 10^{-2}$	1.86	$1.38 \cdot 10^{-2}$	1.86	[1]	—

REFERENCES

1. R. K. CANNAN and A. KIBRICK, J. Am. Chem. Soc., 60, 2314 (1938).

2. C. B. MONK, Trans. Farad. Soc., 47, 297 (1951).

3. M. BABTELSKY and J. BAR-GADDA, Bull. Soc. Chim. France, No. 3, 276 (1953); cited in Ref. zh. khim., 16142 (1955).

4. B. BABTELSKY and J. BAR-GADDA, Bull. Soc. Chim. France, No. 3, 276 (1953); cited in Ref. zh. khim., 16141 (1955).

5. C. W. DAVIES, J. Chem. Soc., 277, (1938).

6. J. SCHUBERT and A. LINDENBAUM, J. Am. Chem. Soc., 74, 3529 (1952).

Malonate Complexes   HOOCCH$_2$COOH   CH$_2$(COO)$_2^{2-}$ (Mal^{2-})

Complex ion	Temp. °C	Ionic strength	Method	k	pk	K	pK	References principal	References supplementary
BaMal	—	0.2	pH-potent.	$5.9 \cdot 10^{-2}$	1.23	$5.9 \cdot 10^{-2}$	1.23	[1]	[3]
CaMal	—	0.2	"	$3.46 \cdot 10^{-2}$	1.46	$3.46 \cdot 10^{-2}$	1.46	[1]	[3, 8]
CdMal	25	0	El.cond.	$5.1 \cdot 10^{-4}$	3.29	$5.1 \cdot 10^{-4}$	3.29	[2]	[3, 10]
CoMal	—	0.04	pH-color.	$1.9 \cdot 10^{-4}$	3.72	$1.9 \cdot 10^{-4}$	3.72	[3]	—
Co(Mal)$_2^{2-}$	—	0	Spectr.	—	—	$7.28 \cdot 10^{-4}$	3.14	[4]	—
CuMal	25	0	El.cond.	$2.5 \cdot 10^{-6}$	5.60	$2.5 \cdot 10^{-6}$	5.60	[2]	[3, 9, 10]
Cu(Mal)$_2^{2-}$	Room	0.03	pH-potent.	—	—	$5.4 \cdot 10^{-8}$	7.27	[5]	[11]
FeMal	—	1.0	Polarog.	$1.58 \cdot 10^{-3}$	2.8	$1.58 \cdot 10^{-3}$	2.8	[6]	—
Fe(Mal)$_2^{2-}$	25	0.5	"	—	—	$6 \cdot 10^{-3}$	2.22	[7]	—
Fe(Mal)$_3^{3-}$	25	0.5	"	—	—	$2.2 \cdot 10^{-16}$	15.66	[7]	—
MgMal	25	0	El.cond.	$1.6 \cdot 10^{-3}$	2.80	$1.6 \cdot 10^{-3}$	2.80	[2]	[1, 3, 10]
MnMal	—	0.04	pH-color.	$5.1 \cdot 10^{-4}$	3.29	$5.1 \cdot 10^{-4}$	3.29	[3]	—
NiMal	18	0	El.cond.	$7.3 \cdot 10^{-5}$	4.14	$7.3 \cdot 10^{-5}$	4.14	[2]	[3, 9]
SrMal	25	0.16	Ion exch.	$5.37 \cdot 10^{-2}$	1.27	$5.37 \cdot 10^{-2}$	1.27	[8]	[1]
ZnMal	25	—	El.cond.	$4.5 \cdot 10^{-4}$	3.35	$4.5 \cdot 10^{-4}$	3.35	[9]	[1, 2, 10]

REFERENCES

1. R. K. CANNAN and A. KIBRICK, J. Am. Chem. Soc., 60, 2314 (1938).

2. R. W. MONEY and C. W. DAVIES, Trans. Farad. Soc., 28, 609 (1932).

3. D. I. STOCK and C. W. DAVIES, J. Chem. Soc., 1371 (1949).

4. M. BOBTELSKY and J. BAR-GADDA, Bull. Soc. Chim. France, Nos. 7-8, 687 (1953); cited in Ref. zh. khim., 16142 (1955).

5. H. L. RILEY, J. Chem. Soc., 1307 (1929); cited in Zbl., 1, 957 (1930).

6. A. E. MARTELL and M. CALVIN, Chemistry of the Metal Chelate Compounds, New York (1953).

7. W. B. SCHAAP, H. A. LAITINEN and J. C. BAILAR, J. Am. Chem. Soc., 76, 5868 (1954).

8. J. SCHUBERT and A. LINDENBAUM, ibid., 74, 3529 (1952).

9. D. J. G. IVES and H. L. RILEY, J. Chem. Soc., 1998 (1931).

10. H. L. RILEY and N. J. FISCHER, ibid., 2006 (1929); cited in Zbl., II, 3110 (1929).

11. M. BOBTELSKY and J. BAR-GADDA, Bull. Soc. Chim. France, No. 3, 276 (1953); cited in Ref. zh. khim., 16141 (1955).

Malate Complexes $^-$OOCCHOHCH$_2$COO$^-$(Ap^{2-})

Complex ion	Temp. °C	Ionic strength	Method	k	pk	K	pK	Refs. prin.	sup.
BaAp	—	0.2	pH-potent	$5 \cdot 10^{-2}$	1.30	$5 \cdot 10^{-2}$	1.30	[1]	[2]
CaAp	—	0.2	„	$1.6 \cdot 10^{-2}$	1.80	$1.6 \cdot 10^{-2}$	1.80	[1]	[2,3]
MgAp	—	0.2	„	$2.8 \cdot 10^{-2}$	1.55	$2.8 \cdot 10^{-2}$	1.55	[1]	[2]
SrAp	—	0.2	„	$3.5 \cdot 10^{-2}$	1.45	$3.5 \cdot 10^{-2}$	1.45	[1]	[3]
ZnAp	—	0.2	„	$1.6 \cdot 10^{-3}$	2.80	$1.6 \cdot 10^{-3}$	2.80	[1]	[2]

REFERENCES

1. R.K.CANNAN and A.KIBRICK, J. Am. Chem. Soc., 60, 2314 (1938).
2. C.B.MONK, Trans. Farad. Soc., 47, 297 (1951).
3. J. SCHUBERT and A. LINDENBAUM, J. Am. Chem. Soc., 74, 3529 (1952).

Nitroacetate Complexes O$_2$NCH$_2$COO$^-$ (Nac$^-$)

Complex ion	Temp. °C	Ionic strength	Method	k	pk	K	pK
AlNac^{2+}	18	0.6	Kin.	$3.3 \cdot 10^{-1}$	0.48	$3.3 \cdot 10^{-1}$	0.48
BeNac$^+$	18	0.6	„	$5.5 \cdot 10^{-1}$	0.26	$5.5 \cdot 10^{-1}$	0.26
CaNac$^+$	18	0.6	„	2.0	−0.30	2.0	−0.30
CdNac$^+$	18	0.6	„	$6.5 \cdot 10^{-1}$	0.19	$6.5 \cdot 10^{-1}$	0.19
CoNac$^+$	18	0.6	„	1.0	0.0	1.0	0.0
CuNac$^+$	18	0.6	„	$3.6 \cdot 10^{-1}$	0.44	$3.6 \cdot 10^{-1}$	0.44
MgNac$^+$	18	0.6	„	1.55	−0.19	1.55	−0.19
NiNac$^+$	18	0.6	„	$8.7 \cdot 10^{-1}$	0.06	$8.7 \cdot 10^{-1}$	0.06
PbNac$^+$	18	0.6	„	$7.2 \cdot 10^{-1}$	0.14	$7.2 \cdot 10^{-1}$	0.14
ZnNac$^+$	18	0 6	„	$9.3 \cdot 10^{-1}$	0.03	$9.3 \cdot 10^{-1}$	0.03

REFERENCE

1. K. J. PEDERSEN, Acta Chem. Scand., 3, 656 (1949); cited in Chem. Abs., 44, 2341 (1950).

Oxalate Complexes $C_2O_4^{2-}$ (Ox^{2-})

Complex ion	Temp. °C	Ionic strength	Method	k	pk	K	pK	References principal	References supplementary
$Al(Ox)_2^-$	—	>0.01	pH-potent.	—	—	$1 \cdot 10^{-13}$	13.0	[1]	
$Al(Ox)_3^{3-}$	—	>0.01	"	$1.6 \cdot 10^{-4}$	3.8	$1.6 \cdot 10^{-17}$	16.8	[1]	
$BaOx$	18	0	El.cond.	$4.7 \cdot 10^{-3}$	2.31	$4.7 \cdot 10^{-3}$	2.31	[2]	
$CaOx$	18	0	"	$1.0 \cdot 10^{-3}$	3.0	$1.0 \cdot 10^{-3}$	3.0	[2]	
$CdOx$	25	0	Solub.	$3.0 \cdot 10^{-4}$	3.52	$3.0 \cdot 10^{-4}$	3.52	[3]	[2]
$Cd(Ox)_2^{2-}$	25	0	.	$1.4 \cdot 10^{-2}$	1.85	$4.2 \cdot 10^{-6}$	5.37	[3]	[9]
$CeOx^+$	25	0	.	$3.0 \cdot 10^{-7}$	6.52	$3.0 \cdot 10^{-7}$	6.52	[3]	
$Ce(Ox)_2^-$	25	0	.	$1.1 \cdot 10^{-4}$	3.96	$3.3 \cdot 10^{-11}$	10.5	[4]	
$Ce(Ox)_3^{3-}$	25	0	.	$1.5 \cdot 10^{-1}$	0.82	$5.0 \cdot 10^{-12}$	11.3	[4]	
$CoOx$	18	0	El.cond.	$2.0 \cdot 10^{-5}$	4.70	$2.0 \cdot 10^{-5}$	4.70	[4]	
$Co(Ox)_2^{2-}$	25	>0.1	Thermodyn.	$3.9 \cdot 10^{-3}$	2.41	$7.8 \cdot 10^{-8}$	7.11	[2]	[9]
$Co(Ox)_3^{4-}$	25	0.1	Solub.	—	—	$1.1 \cdot 10^{-8}$	7.96	[5]	
$CuOx$	18	0	El.cond.	$7.0 \cdot 10^{-7}$	6.16	$7.0 \cdot 10^{-7}$	6.16	[5]	
$Cu(Ox)_2^{2-}$	25	>0.1	Thermodyn.	—	—	$9.1 \cdot 10^{-9}$	8.04	[6]	[13]
$FeOx$	18	0	El.cond.	$2 \cdot 10^{-4}$	4.7	$2.0 \cdot 10^{-4}$	4.7	[2]	
$Fe(Ox)_2^{2-}$	25	0.5	Polarog.	—	—	$3.0 \cdot 10^{-5}$	4.52	[7]	[5]

Oxalate Complexes (contd.)

Complex ion	Temp. °C	Ionic strength	Method	a	pk	K	pK	References prin-cipal	References supple-mentary
$Fe(Ox)_3^{4-}$	5	0.5	Polarog.	$2 \cdot 10^{-1}$	0.70	$6 \cdot 10^{-6}$	5.22	[7]	—
$FeOx^+$	—	—	pH-potent.	$4.0 \cdot 10^{-10}$	9.4	$4.0 \cdot 10^{-10}$	9.4	[8]	—
$Fe(Ox)_2^-$	—	—		$1.6 \cdot 10^{-7}$	6.8	$8.3 \cdot 10^{-17}$	16.2	[8]	—
$Fe(Ox)_3^{3-}$	18	0	pH-potent.	$1 \cdot 10^{-4}$	4.0	$6.3 \cdot 10^{-21}$	20.2	[8]	[7]
$MgOx$	25	0.02	El.cond.	$3.7 \cdot 10^{-4}$	3.43	$3.7 \cdot 10^{-4}$	3.43	[2]	[14]
$Mg(Ox)_2^{2-}$	18	0	Solub.	—	—	$4.2 \cdot 10^{-5}$	4.38	[9]	—
$MnOx$	25	>0.1	El.cond.	$1.3 \cdot 10^{-4}$	3.89	$1.3 \cdot 10^{-4}$	3.89	[2]	—
$Mn(Ox)_2^{2-}$	25.2	2.0	Thermodyn.	—	—	$1.6 \cdot 10^{-6}$	5.80	[5]	—
$MnOx^+$	25.2	2.0	Kin.	$1.05 \cdot 10^{-10}$	9.98	$1.05 \cdot 10^{-10}$	9.98	[10]	—
$Mn(Ox)_2^-$	25.2	2.0		$2.6 \cdot 10^{-7}$	6.59	$2.72 \cdot 10^{-17}$	16.57	[10]	—
$Mn(Ox)_3^{3-}$	25	0	Solub.	$1.4 \cdot 10^{-3}$	2.85	$3.82 \cdot 10^{-20}$	19.42	[10]	—
$NdOx^+$	25	0		$6.2 \cdot 10^{-8}$	7.21	$6.2 \cdot 10^{-8}$	7.21	[4]	—
$Nd(Ox)_2$	18	0	El.cond.	$5.0 \cdot 10^{-5}$	4.3	$3.1 \cdot 10^{-12}$	11.5	[4]	—
$NiOx$	25	0	Thermodyn.	$5 \cdot 10^{-6}$	5.3	$5 \cdot 10^{-6}$	5.3	[2]	—
$Ni(Ox)_2^-$	25	>0.1	Spectr.	—	—	$2.3 \cdot 10^{-8}$	7.64	[5]	[9]
NpO_2Ox^+	25	0.5		$5 \cdot 10^{-4}$	3.30	$5 \cdot 10^{-4}$	3.30	[11]	—
$NpO_2(Ox)_2^{3-}$	25	0.5	El.cond.	$1.7 \cdot 10^{-4}$	3.77	$8.5 \cdot 10^{-8}$	7.07	[11]	—
$SrOx$	18	0	pH-potent.	$2.9 \cdot 10^{-3}$	2.54	$2.9 \cdot 10^{-3}$	2.54	[2]	—
$Th(Ox)_4^{4-}$	30	—	Solub.	—	—	$3.3 \cdot 10^{-26}$	24.48	[12]	—
$YbOx^+$	25	0		$5.0 \cdot 10^{-8}$	7.30	$5.0 \cdot 10^{-8}$	7.30	[4]	—
$Yb(Ox)_2^-$	25	0	El.cond.	$2.6 \cdot 10^{-5}$	4.41	$2.0 \cdot 10^{-12}$	11.7	[4]	—
$ZnOx$	18	0	Thermodyn.	$1.3 \cdot 10^{-5}$	4.89	$1.3 \cdot 10^{-5}$	4.89	[2]	—
$Zn(Ox)_2^{2-}$	25	>0.1		—	—	$2.5 \cdot 10^{-8}$	7.60	[5]	[3]

REFERENCES

1. S. LACROIX, Bull. Soc. Chim. France, 408 (1947); cited in Chem. Abs., 42, 1842 (1948).

2. R. W. MONEY and C. W. DAVIES, Trans. Farad. Soc., 28, 609 (1932).

3. W. J. CLAYTON and W. C. VOSBURGH, J. Am. Chem. Soc., 59, 2414 (1937).

4. C. E. CROUTHAMEL and D. S. MARTIN, Jr., ibid., 73, 569 (1951).

5. E. K. ZOLOTAREV, The Study of Oxalate Complexes in Solution (Izucheniye oksalatnykh kompleksov v rastvore), Dissertatsiya, Khimiko-technolog. inst., Ivanova (1956).

6. J. M. PEACOCK and J. C. JAMES, J. Chem. Soc., 2233 (1951).

7. W. B. SCHAAP, H. A. LAITINEN and J. C. BAILAR, Jr., J. Am. Chem. Soc., 76, 5868 (1954).

8. J. BADOZ-LAMBLING, Ann. chim., 8, No. 12, 586 (1953); cited in Ref. zh. khim., 13790 (1955).

9. E. BARNEY, W. J. ARGERSINGER and C. A. REYNOLDS, J. Am. Chem. Soc., 73, 3785 (1951).

10. H. TAUBE, ibid., 70, 3928 (1948).

11. D. M. GRUEN and J. J. KATZ, J. Am. Chem. Soc., 75, 3772 (1953).

12. M. BOSE and D. M. CHOWDHURY, J. Indian Chem. Soc., 31, No. 2, 111 (1954); cited in Ref. zh. khim., 16166 (1955).

13. H. T. S. BRITTON and M. E. JARRET, J. Chem. Soc., 1498 (1936).

14. R. K. CANNAN and A. KIBRICK, J. Am. Chem. Soc., 60, 2314 (1938).

Oxalacetate Complexes

$$COO^-$$
$$|$$
$$CO$$
$$|$$ $$(Oxac^{2-})$$
$$CH_2$$
$$|$$
$$COO^-$$

Complex ion	Temp. °C	Ionic strength	Method	k	pk	K	pK
DyOxac$^+$	25	0	pH-potent.	$2.2 \cdot 10^{-6}$	5.66	$2.2 \cdot 10^{-6}$	5.66
Dy(Oxac)$_2^-$	25	0	.	$3.3 \cdot 10^{-5}$	4.48	$7.2 \cdot 10^{-11}$	10.14
GdOxac$^+$	25	0	.	$2.9 \cdot 10^{-6}$	5.54	$2.9 \cdot 10^{-6}$	5.54
Gd(Oxac)$_2^-$	25	0	.	$3.0 \cdot 10^{-5}$	4.53	$8.5 \cdot 10^{-11}$	10.07
LaOxac$^+$	25	0	.	$5.6 \cdot 10^{-6}$	5.25	$5.6 \cdot 10^{-6}$	5.25
LuOxac$^+$	25	0	.	$1.35 \cdot 10^{-6}$	5.87	$1.35 \cdot 10^{-6}$	5.87
Lu(Oxac)$_2^-$	25	0	.	$1.9 \cdot 10^{-5}$	4.72	$2.6 \cdot 10^{-11}$	10.59
Y Oxac$^+$	25	0	.	$2.34 \cdot 10^{-6}$	5.63	$2.34 \cdot 10^{-6}$	5.63
Y(Oxac)$_2^-$	25	0	.	$6.3 \cdot 10^{-5}$	4.20	$1.5 \cdot 10^{-10}$	9.83

REFERENCE

1. E. GELLES and G. H. NANCOLLAS, Trans. Farad. Soc., 52, 98 (1956).

Phthalate Complexes $\langle \rangle \begin{matrix} COO^- \\ COO^- \end{matrix}$ **(Pht²⁻)**

Complex ion	Temp. °C	Ionic strength	Method	k	pk	K	pK	References principal	supplementary
BaPht	25	0.15	Potent.	$1.2 \cdot 10^{-1}$	0.92	$1\ 2 \cdot 10^{-1}$	0.92	[1]	—
CaPht	25	0.15	Potent.	$8.5 \cdot 10^{-2}$	1.07	$8.5 \cdot 10^{-2}$	1.07	[1]	—
CoPht	—	—	Spectr.	$1.55 \cdot 10^{-2}$	1.81	$1.55 \cdot 10^{-2}$	1.81	[2]	—
Co(Pht)$_2^{2-}$	Room	0.03	Potent.	—	—	$3.1 \cdot 10^{-5}$	4.51	[3]	[4]

REFERENCES

1. N. R. JOSEPH, J. Biol. Chem., 164, 529 (1946); cited in Chem. Abs., 40, 6321 (1946).

2. M. BOBTELSKY and J. BAR-GADDA, Bull. Soc. Chim. France, Nos. 7-8, 687 (1953); cited in Ref. zh. khim., 16142 (1955).

3. H. L. RILEY, J. Chem. Soc., 1307 (1929); cited in Zbl., II, 957 (1930).

4. M. BOBTELSKY and J. BAR-GADDA, Bull. Soc. Chim. France, No. 3, 276 (1953); cited in Ref. zh. khim., 16141 (1955).

Salicylate Complexes $C_6H_4(COO)O^{2-}$ (Sal^{2-})

Complex ion	Temp. °C	Ionic strength	Method	k	pk	K	pK	References principal	References supplementary
$AlSal^+$	—	—	Spectr.	$8 \cdot 10^{-15}$	14.10	$8 \cdot 10^{-15}$	14.10	[1]	—
$CaSal$	25	0.15	Potent.	$7.25 \cdot 10^{-1}$	0.14	$7.25 \cdot 10^{-1}$	0.14	[2]	[6]
$CuSal$	—	—	Spectr.	$2.3 \cdot 10^{-11}$	10.6	$2.3 \cdot 10^{-11}$	10.6	[3]	—
$Cu(Sal)_2^{2-}$	—	—	"	$5 \cdot 10^{-7}$	6.3	$1.26 \cdot 10^{-17}$	16 9	[3]	[7]
$FeSal^+$	—	—	"	$4.0 \cdot 10^{-17}$	16.4	$4.0 \cdot 10^{-17}$	16.4	[4]	—
$Fe(Sal)_2^-$	—	—	"	$3.5 \cdot 10^{-12}$	11.46	$1.4 \cdot 10^{-28}$	27.85	[4]	—
$Fe(Sal)_3^{3-}$	—	—	"	$2 \cdot 10^{-6}$	5.7	$2.8 \cdot 10^{-34}$	33.55	[4]	—
UO_2Sal	—	—	"	$4 \cdot 10^{-14}$	13.4	$4 \cdot 10^{-14}$	13.4	[5]	—

REFERENCES

1. A. K. BABKO and T. N. RYCHKOVA, Zh. obshch. khim., 18, 1617 (1948).
2. N. R. JOSEPH, J. Biol. Chem., 164, 529 (1946); cited in Chem. Abs., 40, 6321 (1946).
3. A. K. BABKO, Zh. obshch. khim., 17, 443 (1947).
4. A. K. BABKO, ibid., 15, 745 (1945).
5. A. K. BABKO and L. S. KOTELYANSKAYA, Khimsbornik Kievskogo Gosuniversiteta, No. 5, 75 (1949).
6. C. W. DAVIES, J. Chem. Soc., 277 (1938).
7. M. BOBTELSKY and J. BAR-GADDA, Bull. Soc. Chim. France, No. 3, 276 (1953); cited in Ref. zh. khim., 16141 (1955).

Succinate Complexes $\begin{array}{c} CH_2COO^- \\ | \\ CH_2COO^- \end{array}$ (Suc^{2-})

Complex ion	Temp. °C	Ionic strength	Method	k	pk	K	pK	References	
								principal	supplementary
BaSuc	25	0.15	Potent.	$1.07 \cdot 10^{-1}$	0.97	$1.07 \cdot 10^{-1}$	0.97	[1]	[3,5]
CaSuc	25	0.15	"	$6.9 \cdot 10^{-2}$	1.16	$6.9 \cdot 10^{-2}$	1.16	[1]	[3,5,6]
CoSuc	—	—	Spectr.	$7.2 \cdot 10^{-1}$	0.14	$7.2 \cdot 10^{-1}$	0.14	[2]	—
MgSuc	—	0.2	pH-potent.	$6.3 \cdot 10^{-2}$	1.20	$6.3 \cdot 10^{-2}$	1.20	[3]	[5]
RaSuc	25	0.02	Ion exch.	$1 \cdot 10^{-1}$	1.0	$1 \cdot 10^{-1}$	1.0	[4]	—
SrSuc	—	0.2	pH-potent.	$8.7 \cdot 10^{-2}$	1.06	$8.7 \cdot 10^{-2}$	1.06	[3]	[6]
ZnSuc	—	0.2	"	$1.66 \cdot 10^{-2}$	1.78	$1.66 \cdot 10^{-2}$	1.78	[3]	[5]

REFERENCES

1. N. R. JOSEPH, J. Biol. Chem., 164, 529 (1946).

2. M. BOBTELSKY and J. BAR-GADDA, Bull. Soc. Chim. France, Nos. 7-8, 687 (1953); cited in Ref. zh. khim., 16142 (1955).

3. R. K. CANNAN and A. KIBRICK, J. Am. Chem. Soc., 60, 2314 (1938).

4. A. E. MARTELL and M. CALVIN, Chemistry of the Metal Chelate Compounds, New York (1953).

5. C. B. MONK, Trans. Farad. Soc., 47, 297 (1951).

6. J. SCHUBERT and A. LINDENBAUM, J. Am. Chem. Soc., 74, 3529 (1952).

Tartrate Complexes -OOCCHOHCHOHCOO-(Tart²⁻)

Complex ion	Temp. °C	Ionic strength	Method	k	pk	K	pK	References principal	References supplementary
BaTart	—	0.2	pH-potent.	$2.4 \cdot 10^{-2}$	1.62	$2.4 \cdot 10^{-2}$	1.62	[1]	[9]
Bi(HTart)₄⁻	—	—	pH-potent.	—	—	$5 \cdot 10^{-9}$	8.30	[2]	—
Bi(OH)₃Tart²⁻	—	—	"	—	—	$1 \cdot 10^{-31}$	31.0	[2]	—
CaTart	25	0.16	Ion exch.	$1.66 \cdot 10^{-2}$	1.78	$1.66 \cdot 10^{-2}$	1.78	[3]	[1]
CuTart	25	—	Potent.	$9.3 \cdot 10^{-4}$	3.03	$9.3 \cdot 10^{-4}$	3.03	[4]	[10]
Cu(HTart)₂	—	—	Spectr.	—	—	$5 \cdot 10^{-4}$	3.3	[5]	—
Cu(OH)₂(Tart)₂⁴⁻	25	>1.0	Polarog.	—	—	$1.4 \cdot 10^{-10}$	9.85	[6]	—
Cu(OH)Tart⁻	—	—	Solub.	—	—	$3.6 \cdot 10^{-13}$	12.44	[5]	—
Cu(OH)₃Tart²⁻	—	—	"	—	—	$7.3 \cdot 10^{-20}$	19.14	[5]	—
MgTart	—	0.2	pH-potent.	$4.4 \cdot 10^{-2}$	1.36	$4.4 \cdot 10^{-2}$	1.36	[1]	—
Pb(HTart)₃⁻	—	—	Polarog.	—	—	$2 \cdot 10^{-5}$	4.7	[7]	—
Pb(OH)₂Tart²⁻	—	—	"	—	—	$8 \cdot 10^{-15}$	14.1	[7]	[1,9,11]
SrTart	25	0.16	Ion exch.	$2.56 \cdot 10^{-2}$	1.59	$2.56 \cdot 10^{-2}$	1.59	[3]	[4]
ZnTart	—	0.2	pH-potent.	$2.1 \cdot 10^{-3}$	2.68	$2.1 \cdot 10^{-3}$	2.68	[1]	—
Zn(OH)Tart⁻	18	—	Polarog.	—	—	$2.4 \cdot 10^{-8}$	7.62	[8]	—

REFERENCES

1. R.K.CANNAN and A.KIBRICK, J.Am.Chem.Soc., 60, 2314 (1938).
2. A.S.TIKHONOV, Zh.obshch.khim., 24, 37 (1954).
3. J.SCHUBERT and A.LINDENBAUM, J.Am.Chem.Soc., 74, 3529 (1952).
4. SUZUKI, J.Chem.Soc.Japan, Pure Chem.Sect., 72, 524 (1951); cited in Chem.Abs., 46, 3444 (1952).
5. A.S.TIKHONOV and V.P.BEL'SKAYA, Collected Articles on General Chemistry (Sbornik statei po obshchei khimii), II, 1211 (1953).
6. L.MEITES, J.Am.Chem.Soc., 71, 3269 (1949).
7. A.S.TIKHONOV, Trud.Voronezhsk.Universiteta, 32, 113 (1953); cited in Ref. zh.khim., 20995 (1955).
8. N.K.VITCHENKO and A.S.TIKHONOV, Trud.Voronezhsk.Universiteta, 32, 129 (1953); cited in Ref. zh.khim., 20994 (1955).
9. N.R.JOSEPH, J.Biol.Chem., 164, 529 (1946); cited in Chem.Abs., 40, 6321 (1946).
10. R.N.SEN SARMA, J. Indian Chem.Soc., 27, 683 (1950); cited in Chem.Abs., 45, 7907 (1951).
11. J.SCHUBERT and J.W.RICHTER, J.Phys.Colloid Chem., 52, 350 (1948); cited in Chem.Abs., 43, 5301 (1948).

Propionate Complexes
$C_2H_5COO^-$ (Pr⁻)

Complex ion	Temp. °C	Ionic strength	Method	k	pk	K	pK
BaPr⁺	—	0.2	pH-potent.	$4.07 \cdot 10^{-1}$	0.39	$4.07 \cdot 10^{-1}$	0.39
CaPr⁺	—	0.2	"	$2.95 \cdot 10^{-1}$	0.53	$2.95 \cdot 10^{-1}$	0.53
MgPr⁺	—	0.2	"	$3.10 \cdot 10^{-1}$	0.51	$3.10 \cdot 10^{-1}$	0.51
SrPr⁺	—	0.2	"	$3.72 \cdot 10^{-1}$	0.43	$3.72 , 10^{-1}$	0.43
ZnPr⁺	—	0.2	"	$9.35 \cdot 10^{-2}$	1.03	$9.35 \cdot 10^{-2}$	1.03

REFERENCE

1. R. K. CANNAN and A. KIBRICK, J. Am. Chem. Soc., 60, 2314 (1938).

3. Aminoacids

Complexes with Alanine $CH_3(NH_2)CHCOO^-$ (Alan⁻)

Complex ion	Temp. °C	Ionic strength	Method	k	pk	K	pK	References prin-cipalmentary	supplementary
Ag Alan	25	0	Solub.	$2.3 \cdot 10^{-4}$	3.64	$2.3 \cdot 10^{-4}$	3.64	[1]	[5]
Ag(Alan)₂	25	0	"	$2.7 \cdot 10^{-4}$	3.57	$6.2 \cdot 10^{-8}$	7.21	[1]	—
Ba Alan⁺	25	0	—	$1.7 \cdot 10^{-1}$	0.77	$1.7 \cdot 10^{-1}$	0.77	[1]	—
Ca Alan⁺	25	0	El.cond.	$5.75 \cdot 10^{-2}$	1.24	$5.75 \cdot 10^{-2}$	1.24	[2]	—
Co Alan⁺	25	0	pH-potent.	$1.51 \cdot 10^{-5}$	4.82	$1.51 \cdot 10^{-5}$	4.82	[1]	[4]
Co (Alan)₂	25	0	"	$2.2 \cdot 10^{-4}$	3.66	$3.3 \cdot 10^{-9}$	8.48	[1]	[6]
Cu Alan⁺	25	0	"	$3.1 \cdot 10^{-9}$	8.51	$3.1 \cdot 10^{-9}$	8.51	[1]	
Cu (Alan)₂	25	0	"	$1.35 \cdot 10^{-7}$	6.87	$4.2 \cdot 10^{-16}$	15.38	[1]	[4, 6, 7]
Fe (Alan)₂	20	0 01	"	—	—	$5.0 \cdot 10^{-8}$	7.3	[3]	—
Mg Alan⁺	25	0	"	$1.1 \cdot 10^{-2}$	1.96	$1.1 \cdot 10^{-2}$	1.96	[1]	—
Mn Alan⁺	25	0	"	$9.5 \cdot 10^{-4}$	3.02	$9.5 \cdot 10^{-4}$	3.02	[1]	—
Mn (Alan)₂	25	—	"	$9.3 \cdot 10^{-4}$	3.03	$8.9 \cdot 10^{-7}$	6.05	[4]	—
Ni Alan⁺	25	0	Solub.	$1.1 \cdot 10^{-6}$	5.96	$1.1 \cdot 10^{-6}$	5.96	[1]	—
Ni (Alan)₂	25	0	"	$2.0 \cdot 10^{-5}$	4.70	$2.2 \cdot 10^{-11}$	10.66	[1]	—
Pb Alan⁺	25	0	pH-potent.	$1.0 \cdot 10^{-5}$	5.00	$1 0 \cdot 10^{-5}$	5.00	[1]	[5]
Pb (Alan)₂	25	0	"	$5.75 \cdot 10^{-4}$	3.24	$5.75 \cdot 10^{-9}$	8.24	[1]	—
Zn Alan⁺	25	0	"	$6.2 \cdot 10^{-6}$	5 21	$6.2 \cdot 10^{-6}$	5.21	[1]	—
Zn (Alan)₂	25	0	"	$4.7 \cdot 10^{-5}$	4.33	$2.9 \cdot 10^{-10}$	9 54	[1]	—

REFERENCES

1. C. B. MONK, Trans. Farad. Soc., 47, 285, 292, 297 (1951).
2. C. W. DAVIES and G. M. WAIND, J. Chem. Soc., 305 (1950).
3. A. E. MARTELL and M. CALVIN, Chemistry of the Metal Chelate Compounds, New York (1953).
4. L. E. MALEV and D. P. MELLOR, Nature, 165, 453 (1950).
5. R. M. KEEFER and H. R. REIBER, J. Am. Chem. Soc., 63, 689 (1941).
6. R. M. KEEFER, ibid., 68, 2329 (1946); 70, 476 (1948).
7. N. C. LI and E. DOODY, J. Am. Chem. Soc., 72, 1891 (1950).

Complexes with 2-Sulphoanilinediacetic Acid

$$\begin{array}{c} SO_3^- \\ | \\ \langle\ \rangle - N \begin{cases} CH_2COO^- \\ CH_2COO^- \end{cases} \end{array} \quad (Saa^{3-})$$

Complex ion	Temp. °C	Ionic strength	Method	k	pk	K	pK
BaSaa⁻	20	0.1	pH-potent.	$5.5 \cdot 10^{-3}$	2.26	$5.5 \cdot 10^{-3}$	2.26
CaSaa⁻	20	0.1	"	$2.7 \cdot 10^{-5}$	4.57	$2.7 \cdot 10^{-5}$	4.57
LiSaa²⁻	20	0.1	"	$5.5 \cdot 10^{-3}$	2.26	$5.5 \cdot 10^{-3}$	2.26
MgSaa⁻	20	0.1	"	$2.2 \cdot 10^{-3}$	2.68	$2.2 \cdot 10^{-3}$	2.68
NaSaa²⁻	20	0.1	"	$1.1 \cdot 10^{-1}$	0.98	$1.1 \cdot 10^{-1}$	0.98
SrSaa⁻	20	0.1	"	$3.2 \cdot 10^{-1}$	3.50	$3.2 \cdot 10^{-4}$	3.50

REFERENCE

1. G. SCHWARZENBACH, A. WILLI and R. O. BACH, Helv. chim. Acta, 30, 1303 (1947).

Complexes with Aminobarbituric-N,N-Diacetic Acid (Amac³⁻)

$$N = C - O^- \quad CH_2COO^-$$
$$CO \qquad Ch - N$$
$$NH - CO \qquad CH_2COO^-$$

Complex ion	Temp. °C	Ionic strength	Method	k	pk	K	pK	References
Ba Amac⁻	20	0	pH-potent.	$1.65 \cdot 10^{-7}$	6.78	$1.65 \cdot 10^{-7}$	6.78	[1]
Ca Amac⁻	20	0	•	$1.70 \cdot 10^{-9}$	8.77	$1.70 \cdot 10^{-9}$	8.77	[1]
Ca (Amac)₂⁴⁻	Room	~0.01	•	$6.3 \cdot 10^{-6}$	5.2	—	—	[2]
Cd (Amac)₂⁴⁻	•	~0.01	•	$2.0 \cdot 10^{-7}$	6.7	—	—	[2]
Ce (Amac)₃³⁻	•	~0.01	•	10^{-10}	10	—	—	[2]
Co (Amac)₂⁴⁻	•	~0.01	•	$6.3 \cdot 10^{-4}$	3.2	—	—	[2]
La (Amac)₂³⁻	•	~0.01	•	10^{-10}	10	—	—	[2]
Li Amac²⁻	20	0	•	$4.00 \cdot 10^{-6}$	5.40	$4.00 \cdot 10^{-6}$	5.40	[1]
Mg Amac⁻	20	0	•	$1.45 \cdot 10^{-9}$	8.84	$1.45 \cdot 10^{-9}$	8.84	[1]
Mg (Amac)₂⁴⁻	Room	~0.01	•	$8.0 \cdot 10^{-4}$	3.1	$8.0 \cdot 10^{-4}$	3.1	[1]
Mn (Amac)₂⁴⁻	•	~0.01	•	$1.0 \cdot 10^{-4}$	4.0	—	—	[2]
Na Amac³⁻	20	0	•	$4.80 \cdot 10^{-4}$	3.32	$4.80 \cdot 10^{-4}$	3.32	[1]
Ni (Amac)₂⁴⁻	Room	~0.01	•	$5.0 \cdot 10^{-4}$	3.3	—	—	[2]
Sr Amac⁻	20	0	•	$2.24 \cdot 10^{-8}$	7.65	$2.24 \cdot 10^{-8}$	7.65	[1]
Zn (Amac)₂⁴⁻	Room	~0.01	•	$6.3 \cdot 10^{-4}$	3.2	—	—	[2]

REFERENCES

1. G. SCHWARZENBACH, E. KAMPITSCH and R. STEINER, Helv. chim. Acta, 29, 364 (1946).
2. G. SCHWARZENBACH and W. BIEDERMAN, ibid., 31, 456 (1948).

Complexes with Asparagine

$$NH_2$$
$$|$$
$$H_2NOCCH_2CHCOO^- \quad (Asp^-)$$

Complex ion	Temp. °C	Ionic strength	Method	k	pk	K	pK
Cd (Asp)₂	20	0.01	pH-potent.	—	—	$1 6 \cdot 10^{-7}$	6.8
Co (Asp)₂	20	0.01	"	—	—	$4.0 \cdot 10^{-9}$	8.4
Cu (Asp)₂	20	0.01	"	—	—	$1.25 \cdot 10^{-15}$	14.9
Fe (Asp)₂	20	0.01	"	—	—	$3.16 \cdot 10^{-7}$	6.5
Mg (Asp)₂	20	0.01	"	—	—	$1 \cdot 10^{-4}$	4.0
Mn (Asp)₂	20	0.01	"	—	—	$3 \cdot 10^{-5}$	4.5
Ni (Asp)₂	20	0.01	"	—	—	$2.5 \cdot 10^{-11}$	10.6
Zn (Asp)₂	20	0.01	"	—	—	$2.0 \cdot 10^{-9}$	8.7

REFERENCE

1. A. E. MARTELL and M. CALVIN, Chemistry of the Metal Chelate Compounds, New York (1953).

Complexes with N-Hydroxyethylethylenediaminetriacetic Acid

$$H_2C - N\begin{array}{c} CH_2COO^- \\ CH_2COO^- \end{array}$$
$$|$$
$$H_2C - N\begin{array}{c} CH_2COO^- \\ CH_2CH_2OH \end{array} \quad (Hed^{3-})$$

Complex ion	Temp. °C	Ionic strength	Method	k	pk	K	pK
CaHed⁻	29.6	0.1	pH-potent.	$1.0 \cdot 10^{-8}$	8.0	$1.0 \cdot 10^{-8}$	8.0
CdHed⁻	29.6	0.1	"	$1.0 \cdot 10^{-13}$	13.0	$1.0 \cdot 10^{-13}$	13.0
CoHed⁻	29.6	0.1	"	$4.0 \cdot 10^{-15}$	14.4	$4.0 \cdot 10^{-15}$	14.4
CuHed⁻	29.6	0.1	"	$4.0 \cdot 10^{-18}$	17.4	$4.0 \cdot 10^{-18}$	17.4
FeHed⁻	29.6	0.1	"	$2.5 \cdot 10^{-12}$	11.6	$2.5 \cdot 10^{-12}$	11.6
MnHed⁻	29.6	0.1	"	$2.0 \cdot 10^{-11}$	10.7	$2.0 \cdot 10^{-11}$	10.7
NiHed⁻	29.6	0.1	"	$1.0 \cdot 10^{-17}$	17.0	$1.0 \cdot 10^{-17}$	17.0
ZnHed⁻	29.6	0.1	"	$3.2 \cdot 10^{-15}$	14.5	$3.2 \cdot 10^{-15}$	14.5

REFERENCE

1. S. CHABERECK, Jr., and A. E. MARTELL, J. Am. Chem. Soc., 77, 1477 (1955).

Complexes with Aspartic Acid $-OOCCH_2CHCOO^-$ NH_2 $(Aspa^{2-})$

Complex ion	Temp °C	Ionic strength	Method	k	pk	K	pK	References principal	References supplementary
BaAspa	25	0.1	pH-potent.	$7.2 \cdot 10^{-2}$	1.14	$7.2 \cdot 10^{-2}$	1.14	[1]	—
CaAspa	25	0.1	.	$2.5 \cdot 10^{-2}$	1.60	$2.5 \cdot 10^{-2}$	1.60	[1]	—
CdAspa	30	0.1	.	$4.3 \cdot 10^{-5}$	4.37	$4.3 \cdot 10^{-5}$	4.37	[2]	—
Cd(Aspa)$_2^{2-}$	30	0.1	.	$7.8 \cdot 10^{-4}$	3.11	$3.3 \cdot 10^{-8}$	7.48	[2]	—
CoAspa	30	0.1	.	$1.26 \cdot 10^{-6}$	5.90	$1.26 \cdot 10^{-6}$	5.90	[2]	—
Co(Aspa)$_2^{2-}$	30	0.1	.	$5.25 \cdot 10^{-5}$	4.28	$6.6 \cdot 10^{-11}$	10.18	[2]	—
CuAspa	30	0.1	.	$2.7 \cdot 10^{-9}$	8.57	$2.7 \cdot 10^{-9}$	8.57	[2]	—
Cu(Aspa)$_2^{2-}$	30	0.1	.	$1.66 \cdot 10^{-7}$	6.78	$4.5 \cdot 10^{-16}$	15.35	[2]	[4]
MgAspa	25	0.1	.	$3.7 \cdot 10^{-3}$	2.43	$3.7 \cdot 10^{-3}$	2.43	[1]	—
MnAspa	—	—	.	$1.26 \cdot 10^{-4}$	3.90	$1.26 \cdot 10^{-4}$	3.90	[1]	—
NiAspa	30	0.1	.	$7.6 \cdot 10^{-8}$	7.12	$7.6 \cdot 10^{-8}$	7.12	[2]	—
Ni(Aspa)$_2^{2-}$	30	0.1	.	$5.4 \cdot 10^{-6}$	5.27	$4.1 \cdot 10^{-13}$	12.39	[2]	—
RaAspa	25	0.02	.	$1.38 \cdot 10^{-1}$	0.86	$1.38 \cdot 10^{-1}$	0.86	[1]	—
SrAspa	25	0.10	.	$3.3 \cdot 10^{-2}$	1.48	$3.3 \cdot 10^{-2}$	1.48	[3]	—
ZnAspa	30	0.1	.	$1.45 \cdot 10^{-6}$	5.84	$1.45 \cdot 10^{-6}$	5.84	[2]	—
Zn(Aspa)$_2^{2-}$	30	0.1	.	$4.9 \cdot 10^{-5}$	4.31	$7.1 \cdot 10^{-11}$	10.15	[2]	—

REFERENCES

1. A. E. MARTELL and M. CALVIN, Chemistry of the Metal Chelate Compounds, New York (1953).
2. S. CHABERECK and A. E. MARTELL, J. Am. Chem. Soc., 74, 6021 (1952).
3. J. SCHUBERT and A. LINDENBAUM, ibid., 74, 3529 (1952).
4. N. C. LI and E. DOODY, ibid., 72, 1891 (1950).

Complexes with 1,2-Diaminocyclohexanetetraacetic Acid

$$C - N \begin{cases} CH_2COO^- \\ CH_2COO^- \end{cases}$$
$$H_2C \quad C - N \begin{cases} CH_2COO^- \\ CH_2COO^- \end{cases} \quad (Data^{4-})$$
$$H_2C \quad CH_2$$
$$H_2C$$

Complex ion	Temp. °C	Ionic strength	Method	k	pk	K	pK
AlData⁻	20	0.1	pH-potent.	$2.34 \cdot 10^{-18}$	17.63	$2.34 \cdot 10^{-18}$	17.63
CaData²⁻	20	0.1	"	$8.32 \cdot 10^{-13}$	12.08	$8.32 \cdot 10^{-13}$	12.08
CdData²⁻	20	0.1	"	$5.89 \cdot 10^{-20}$	19.23	$5.89 \cdot 10^{-20}$	19.23
CeData⁻	20	0.1	"	$1.74 \cdot 10^{-17}$	16.76	$1.74 \cdot 10^{-17}$	16.76
CoData²⁻	20	0.1	"	$1.20 \cdot 10^{-19}$	18.92	$1.20 \cdot 10^{-19}$	18.92
CuData²⁻	20	0.1	"	$5.00 \cdot 10^{-22}$	21.30	$5.00 \cdot 10^{-22}$	21.30
DyData⁻	20	0.1	"	$2.04 \cdot 10^{-20}$	19.69	$2.04 \cdot 10^{-20}$	19.69
ErData⁻	20	0.1	"	$2.09 \cdot 10^{-21}$	20.68	$2.09 \cdot 10^{-21}$	20.68
EuData⁻	20	0.1	"	$2.40 \cdot 10^{-19}$	18.62	$2.40 \cdot 10^{-19}$	18.62
GaData⁻	20	0.1	"	$1.23 \cdot 10^{-23}$	22.91	$1.23 \cdot 10^{-23}$	22.91
GdData⁻	20	0.1	"	$1.70 \cdot 10^{-19}$	18.77	$1.70 \cdot 10^{-19}$	18.77
LaData⁻	20	0.1	"	$5.50 \cdot 10^{-17}$	16.26	$5.50 \cdot 10^{-17}$	16.26
LuData⁻	20	0.1	"	$3.09 \cdot 10^{-22}$	21.51	$3.09 \cdot 10^{-22}$	21.51
MnData²⁻	20	0.1	"	$1.66 \cdot 10^{-17}$	16.78	$1.66 \cdot 10^{-17}$	16.78
NdData⁻	20	0.1	"	$2.09 \cdot 10^{-18}$	17.68	$2.09 \cdot 10^{-18}$	17.68
PbData²⁻	20	0.1	"	$2.09 \cdot 10^{-20}$	19.68	$2.09 \cdot 10^{-20}$	19.68
PrData⁻	20	0.1	"	$4.90 \cdot 10^{-18}$	17.31	$4.90 \cdot 10^{-18}$	17.31
SmData⁻	20	0.1	"	$4.17 \cdot 10^{-19}$	18.38	$4.17 \cdot 10^{-19}$	18.38
TbData⁻	20	0.1	"	$3.16 \cdot 10^{-20}$	19.50	$3.16 \cdot 10^{-20}$	19.50
ThData⁻	20	0.1	"	$1.10 \cdot 10^{-21}$	20.96	$1.10 \cdot 10^{-21}$	20.96
VOData⁻	20	0.1	"	$3.98 \cdot 10^{-20}$	19.40	$3.98 \cdot 10^{-20}$	19.40
YData⁻	20	0.1	"	$7.08 \cdot 10^{-20}$	19.15	$7.08 \cdot 10^{-20}$	19.15
YbData⁻	20	0.1	"	$7.59 \cdot 10^{-22}$	21.12	$7.59 \cdot 10^{-22}$	21.12
ZnData²⁻	20	0.1	"	$2.14 \cdot 10^{-19}$	18.67	$2.14 \cdot 10^{-19}$	18.67

REFERENCE

1. G. SCHWARZENBACH, Helv. chim. Acta, 37, 937 (1954).

Complexes with Ethylenediaminetetraacetic Acid

$$H_2C - N \begin{array}{c} CH_2COO^- \\ CH_2COO^- \end{array} (Edta^{4-})$$
$$H_2C - N \begin{array}{c} CH_2COO^- \\ CH_2COO^- \end{array}$$

Complex ion	Temp. °C	Ionic strength	Method	k	pk	K	pK	References principal	References supplementary
BaEdta²⁻	20	0.1	pH-potent	$1.74 \cdot 10^{-8}$	7.76	$1.74 \cdot 10^{-3}$	7.76	[1]	—
CaEdta²⁻	20	0.1	"	$2.58 \cdot 10^{-11}$	10.59	$2.58 \cdot 10^{-11}$	10.59	[1]	—
CdEdta²⁻	20	0.1	"	$3.3 \cdot 10^{-17}$	16.48	$3.3 \cdot 10^{-17}$	16.48	[2]	—
CeEdta⁻	20	0.1	"	$4.1 \cdot 10^{-16}$	15.39	$4.1 \cdot 10^{-16}$	15.39	[3]	—
CoEdta²⁻	20	0.1	"	$7.9 \cdot 10^{-17}$	16.10	$7.9 \cdot 10^{-17}$	16.10	[2]	—
CuEdta²⁻	20	0.1	"	$1.38 \cdot 10^{-19}$	18.86	$1.38 \cdot 10^{-19}$	18.86	[4]	[2]
DyEdta⁻	20	0.1	"	$2.7 \cdot 10^{-18}$	17.57	$2.7 \cdot 10^{-18}$	17.57	[3]	—
ErEdta⁻	20	0.1	"	$1.05 \cdot 10^{-18}$	17.98	$1.05 \cdot 10^{-18}$	17.98	[3]	—
EuEdta⁻	20	0.1	"	$2.04 \cdot 10^{-17}$	16.69	$2.04 \cdot 10^{-17}$	16.69	[3]	—
FeEdta²⁻	20	0.1	"	$3.54 \cdot 10^{-15}$	14.45	$3.54 \cdot 10^{-15}$	14.45	[4]	[2]
FeEdta⁻	20	0.1	"	$8.0 \cdot 10^{-26}$	25.1	$8.0 \cdot 10^{-26}$	25.1	[5]	—
GdEdta⁻	20	0.1	"	$2.0 \cdot 10^{-17}$	16.70	$2.0 \cdot 10^{-17}$	16.70	[3]	—
HgEdta²⁻	20	0.1	"	$7.1 \cdot 10^{-23}$	22.15	$7.1 \cdot 10^{-23}$	22.15	[6]	—
HoEdta⁻	20	0.1	"	$2.14 \cdot 10^{-18}$	17.67	$2.14 \cdot 10^{-18}$	17.67	[3]	—
LaEdta⁻	20	0.1	"	$1.9 \cdot 10^{-15}$	14.72	$1.9 \cdot 10^{-15}$	14.72	[3]	—
LiEdta³⁻	20	0.1	"	$1.62 \cdot 10^{-3}$	2.79	$1.62 \cdot 10^{-3}$	2.79	[1]	—
LuEdta⁻	20	0.1	"	$8.7 \cdot 10^{-20}$	19.06	$8.7 \cdot 10^{-20}$	19.06	[3]	—

Complexes with Ethylenediaminetetraacetic Acid (contd.)

Complex ion	Temp °C	Ionic strength	Method	k	pk	K	pK	References prin-cipal	References supple-mentary
MgEdta²⁻	20	0.1	pH-potent	$2.04 \cdot 10^{-9}$	8.69	$2.04 \cdot 10^{-9}$	8.69	[1]	—
MnEdta²⁻	20	0.1	"	$3.4 \cdot 10^{-14}$	13.47	$3.4 \cdot 10^{-14}$	13.47	[2]	—
NaEdta³⁻	20	0.1	"	$2.2 \cdot 10^{-2}$	1.66	$2.2 \cdot 10^{-2}$	1.66	[1]	—
NdEdta⁻	20	0.1	"	$8.7 \cdot 10^{-17}$	16.06	$8.7 \cdot 10^{-17}$	16.06	[3]	—
NiEdta²⁻	20	0.1	"	$3.54 \cdot 10^{-19}$	18.45	$3.54 \cdot 10^{-19}$	18.45	[2]	[7]
PbEdta²⁻	20	0.1	"	$6.3 \cdot 10^{-19}$	18.2	$6.3 \cdot 10^{-19}$	18.2	[2]	—
PrEdta⁻	20	0.1	"	$1.78 \cdot 10^{-16}$	15.75	$1.78 \cdot 10^{-16}$	15.75	[3]	—
SmEdta⁻	20	0.1	"	$2.8 \cdot 10^{-17}$	16.55	$2.8 \cdot 10^{-17}$	16.55	[3]	—
SrEdta²⁻	20	0.1	"	$2.34 \cdot 10^{-9}$	8.63	$2.34 \cdot 10^{-9}$	8.63	[1]	—
TbEdta⁻	20	0.1	"	$5.6 \cdot 10^{-18}$	17.25	$5.6 \cdot 10^{-18}$	17.25	[3]	—
TuEdta⁻	20	0.1	"	$2.56 \cdot 10^{-19}$	18.59	$2.56 \cdot 10^{-19}$	18.59	[3]	—
YEdta⁻	20	0.1	"	$4.17 \cdot 10^{-18}$	17.38	$4.17 \cdot 10^{-18}$	17.38	[3]	—
YbEdta⁻	20	0.1	"	$2.1 \cdot 10^{-19}$	18.68	$2.1 \cdot 10^{-19}$	18.68	[3]	—
ZnEdta²⁻	20	0.1	"	$2.63 \cdot 10^{-17}$	16.58	$2.63 \cdot 10^{-17}$	16.58	[4]	[2]

REFERENCES

1. G. SCHWARZENBACH and H. ACKERMANN, Helv. chim. Acta, 30, 1798 (1947).
2. G. SCHWARZENBACH and E. FREITAG, ibid., 34, 1503 (1951).
3. E. J. WHEELWRIGHT, F. H. SPEDDING and G. SCHWARZENBACH, J. Am. Chem. Soc., 75, 4196 (1953).
4. H. ACKERMANN and G. SCHWARZENBACH, Helv. chim. Acta, 32, 1543 (1949).
5. G. SCHWARZENBACH and J. HELLER, ibid., 31, 1029 (1948).
6. T. GOFFART, G. MICHEL and G. DUYCKAERTS, Anal. Chim. Acta, 9, 184 (1953).
7. C. M. COOK and F. A. LONG, J. Am. Chem. Soc., 73, 4119 (1951).

Complexes with Glycine $NH_2CH_2COO^-$ (Gl^-)

Complex ion	Temp °C	Ionic strength	Method	k	pk	K	pK	References principal	References supplementary
AgGl	25	0	Solub.	$3.1 \cdot 10^{-4}$	3.51	$3.1 \cdot 10^{-4}$	3.51	[1]	—
Ag(Gl)$_2^-$	25	0		$4.2 \cdot 10^{-4}$	3.38	$1.3 \cdot 10^{-7}$	6.89	[1]	—
BaGl⁺	25	0	pH-potent.	$1.7 \cdot 10^{-1}$	0.77	$1.7 \cdot 10^{-1}$	0.77	[2]	—
CaGl⁺	25	0	Solub.	$3.7 \cdot 10^{-2}$	1.43	$3.7 \cdot 10^{-2}$	1.43	[3]	[8]
Cd(Gl)$_2$	20	0.01	pH-potent.	—	—	$8.0 \cdot 10^{-9}$	8.1	[4]	—
CoGl⁺	25	0	·	$5.9 \cdot 10^{-6}$	5.23	$5.9 \cdot 10^{-6}$	5.23	[2]	[5]
Co(Gl)$_2$	25	0	·	$9.5 \cdot 10^{-5}$	4.02	$5.6 \cdot 10^{-10}$	9.25	[2]	[5]
Co(Gl)$_3$	26	—	·	—	—	$1.75 \cdot 10^{-11}$	10.76	[5]	—
CuGl⁺	25	0.1	·	$4.15 \cdot 10^{-9}$	8.38	$4.15 \cdot 10^{-9}$	8.38	[6]	[9]
Cu(Gl)$_2$	25	0.1	·	$1.35 \cdot 10^{-7}$	6.87	$5.6 \cdot 10^{-16}$	15.25	[6]	
Cu(Gl)$_3$	25	1.0	·	—	—	$5.4 \cdot 10^{-17}$	16.27	[7]	[7,9,10]
Fe(Gl)$_2$	20	0.01	Polarog.	—	—	$1.6 \cdot 10^{-8}$	7.8	[4]	—
MgGl⁺	25	0	pH-potent.	$3.6 \cdot 10^{-4}$	3.44	$3.6 \cdot 10^{-4}$	3.44	[2]	—
Mg(Gl)$_2$	20	0.04	·	—	—	$1 \cdot 10^{-4}$	4.0	[4]	—
MnGl⁺	25	0	·	$3.6 \cdot 10^{-4}$	3.44	$3.6 \cdot 10^{-4}$	3.44	[2]	—
Mn(Gl)$_2$	25	0.01	·	—	—	$3.2 \cdot 10^{-6}$	5.5	[4]	—
NiGl⁺	25	0.1	·	$1.38 \cdot 10^{-6}$	5.86	$1.38 \cdot 10^{-6}$	5.86	[6]	[2]
Ni(Gl)$_2$	25	0.1	·	$1.65 \cdot 10^{-5}$	4.78	$2.3 \cdot 10^{-11}$	10.64	[6]	[2]
PbGl⁺	25	0	·	$3.4 \cdot 10^{-6}$	5.47	$3.4 \cdot 10^{-6}$	5.47	[2]	—
Pb(Gl)$_2$	25	0	·	$4.1 \cdot 10^{-4}$	3.39	$1.38 \cdot 10^{-9}$	8.86	[2]	—
SrGl⁺	25	0.16	Ion exch.	$2.5 \cdot 10^{-1}$	0.6	$2.5 \cdot 10^{-1}$	0.6	[8]	—
ZnGl⁺	25	0	pH-potent.	$3.0 \cdot 10^{-6}$	5.52	$3.0 \cdot 10^{-6}$	5.52	[2]	—
Zn(Gl)$_2$	25	0	·	$3.6 \cdot 10^{-5}$	4.44	$1.1 \cdot 10^{-10}$	9.96	[2]	—

REFERENCES

1. C. B. MONK, Trans. Farad. Soc., 47, 292 (1951).
2. C. B. MONK, ibid., 47, 297 (1951).
3. C. W. DAVIES and G. M. WAIND, J. Chem. Soc., 301 (1950).
4. A. E. MARTELL and M. CALVIN, Chemistry of the Metal Chelate Compounds, New York (1953).
5. J. B. GILBERT, M. C. OTEY and J. Z. HEARON, J. Am. Chem. Soc., 77, 2599 (1955).
6. F. BASOLO and Yun Ti CHEN, ibid., 76, 953 (1954).
7. R. M. KEEFER, ibid., 68, 2329 (1946).
8. J. SCHUBERT and A. LINDENBAUM, ibid., 47, 285 (1951).
9. C. B. MONK, Trans. Farad. Soc., 47, 285 (1951).
10. N. C. LI and E. DOODY, J. Am. Chem. Soc., 74, 4184 (1952).

Complexes with N,N-Dihydroxyethylglycine

$$\begin{array}{l} HOCH_2CH_2 \\ HOCH_2CH_2 \end{array}\!\!\!\Big\rangle N - CH_2COO^- \ (Dge^-)$$

Complex ion	Temp. °C	Ionic strength	Method	k	pk	K	pK
CdDge⁺	30	0.1	pH-potent.	$1.65 \cdot 10^{-5}$	4.78	$1.65 \cdot 10^{-5}$	4.78
Cd (Dge)₂	30	0.1	"	$4.3 \cdot 10^{-4}$	3.37	$7.1 \cdot 10^{-9}$	8.15
CoDge⁺	30	0.1	"	$5.25 \cdot 10^{-6}$	5.28	$5.25 \cdot 0^{-6}$	5.28
Co (Dge)₂	30	0.1	"	$3.0 \cdot 10^{-4}$	3.52	$1.6 \cdot 10^{-9}$	8.80
CuDge⁺	30	0.1	"	$7.1 \cdot 10^{-9}$	8.15	$7.1 \cdot 10^{-9}$	8.15
Cu (Dge)₂	30	0.1	"	$6.3 \cdot 10^{-6}$	5.20	$4.5 \cdot 10^{-14}$	13.35
FeDge⁺	30	0.1	"	$5.4 \cdot 10^{-5}$	4.27	$5.4 \cdot 10^{-5}$	4.27
Fe (Dge)₂	30	0.1	"	$1.0 \cdot 10^{-3}$	3.00	$5.4 \cdot 10^{-8}$	7.27
MgDge⁺	30	0.1	"	$7.1 \cdot 10^{-2}$	1.15	$7.1 \cdot 10^{-2}$	1.15
MnDge⁺	30	0.1	"	$8.3 \cdot 10^{-4}$	3.08	$8.3 \cdot 10^{-4}$	3.08
Mn (Dge)₂	30	0.1	"	$4.7 \cdot 10^{-3}$	2.33	$3.9 \cdot 10^{-6}$	5.41
NiDge⁺	30	0.1	"	$4.2 \cdot 10^{-7}$	6.38	$4.2 \cdot 10^{-7}$	6.38
Ni (Dge)₂	30	0.1	"	$4.0 \cdot 10^{-5}$	4.40	$1.65 \cdot 10^{-11}$	10.78
ZnDge⁺	30	0.1	"	$4.4 \cdot 10^{-6}$	5.36	$4.4 \cdot 10^{-6}$	5.36
Zn (Dge)₂	30	0.1	"	$5.5 \cdot 10^{-14}$	3.26	$2.4 \cdot 10^{-9}$	8.62

REFERENCE

1. S. CHABERECK, R. C. COURNEY and A. E. MARTELL, J. Am. Chem. Soc., 75, 2185 (1953).

Complexes with Glycyl-glycine $NH_2CH_2CONHCH_2COO^-$ (Glgl$^-$)

Complex ion	Temp. °C	Ionic strength	Method	k	p^k	K	pK	References principal	References supplementary
AgGlgl	25	0	pH-potent,	$1.9 \cdot 10^{-3}$	2.72	$1\ 9 \cdot 10^{-3}$	2.72	[1]	
Ag(Glgl)$_2^-$	25	0	"	$5.5 \cdot 10^{-3}$	2.26	$1.0 \cdot 10^{-5}$	5.0	[1]	
CaGlgl$^+$	25	0	"	$5.7 \cdot 10^{-2}$	1.25	$5.7 \cdot 10^{-2}$	1.25	[2]	
CoGlgl$^+$	25	0	"	$3.24 \cdot 10^{-4}$	3.49	$3.24 \cdot 10^{-4}$	3.49	[3]	[5]
Co(Glgl)$_2$	25	0	"	$4.1 \cdot 10^{-3}$	2.39	$1.32 \cdot 10^{-6}$	5.88	[3]	[5]
CuGlgl$^+$	25	0	"	$9.2 \cdot 10^{-7}$	6.04	$9.2 \cdot 10^{-7}$	6.04	[4]	
Cu(Glgl)$_2$	25	0	"	$2.4 \cdot 10^{-6}$	5.64	$2.2 \cdot 10^{-12}$	11.66	[4]	
MgGlgl$^+$	25	0	"	$8.7 \cdot 10^{-2}$	1.06	$8.7 \cdot 10^{-2}$	1.06	[3]	
MnGlgl$^+$	25	0	"	$7.1 \cdot 10^{-3}$	2.15	$7.1 \cdot 10^{-3}$	2.15	[3]	
NiGlgl$^+$	25	0	"	$3.22 \cdot 10^{-5}$	4.49	$3.22 \cdot 10^{-5}$	4.49	[3]	
Ni(Glgl)$_2$	25	0	"	$3.8 \cdot 10^{-4}$	3.42	$1\ 22 \cdot 10^{-8}$	7.91	[3]	
PbGlgl$^+$	25	0	"	$5.9 \cdot 10^{-4}$	3.23	$5.9 \cdot 10^{-4}$	3.23	[3]	
Pb(Glgl)$_2$	25	0	"	$2.0 \cdot 10^{-3}$	2.70	$1.18 \cdot 10^{-6}$	5.93	[3]	
ZnGlgl$^+$	25	0	"	$1.6 \cdot 10^{-4}$	3.80	$1.6 \cdot 10^{-4}$	3.80	[3]	
Zn(Glgl)$_2$	25	0	"	$1.7 \cdot 10^{-3}$	2.77	$2.7 \cdot 10^{-7}$	6.57	[3]	

REFERENCES

1. C. B. MONK, Trans. Farad. Soc., 47, 292 (1951).
2. C. W. DAVIES and G. M. WAIND, J. Chem. Soc., 301 (1950).
3. C. B. MONK, Trans. Farad. Soc., 47, 297 (1951).
4. C. B. MONK, ibid., 285 (1951).
5. J. B. GILBERT, M. C. OTEY and J. Z. HEARON, J. Am. Chem. Soc., 77, 2599 (1955).

Complexes with β-Hydroxyethyliminodiacetic Acid $HOCH_2CH_2N \genfrac{}{}{0pt}{}{CH_2COO^-}{CH_2COO^-}$ (Himda^{2-})

Complex ion	Temp. °C	Ionic strength	Method	k	pk	K	pK
CaHimda	30	0.1	pH-potent.	$1.48 \cdot 10^{-5}$	4.83	$1.48 \cdot 10^{-5}$	4.83
CdHimda	30	0.1	"	$6.6 \cdot 10^{-8}$	7.12	$6.6 \cdot 10^{-8}$	7.12
Cd(Himda)$_2^{2-}$	30	0.1	"	$7.6 \cdot 10^{-6}$	5.12	$5.6 \cdot 10^{-13}$	12.24
CoHimda	30	0.1	"	$5.4 \cdot 10^{-9}$	8.27	$5.4 \cdot 10^{-9}$	8.27
Co(Himda)$_2^{2-}$	30	0.1	"	$3.6 \cdot 10^{-5}$	4.44	$1.95 \cdot 10^{-13}$	12.71
Cu(Himda)$_2^{2-}$	30	0.1	"	$5.9 \cdot 10^{-5}$	4.23	—	—
MgHimda	30	0.1	"	$2.9 \cdot 10^{-4}$	3.54	$2.9 \cdot 10^{-4}$	3.54
MnHimda	30	0.1	"	$2.24 \cdot 10^{-6}$	5.65	$2.24 \cdot 10^{-6}$	5.65
Mn(Himda)$_2^{2-}$	30	0.1	"	$1.18 \cdot 10^{-4}$	3.93	$2.6 \cdot 10^{-10}$	9.58
NiHimda	30	0.1	"	$2.9 \cdot 10^{-10}$	9.54	$2.9 \cdot 10^{-10}$	9.54
Ni(Himda)$_2^{2-}$	30	0.1	"	$7.1 \cdot 10^{-6}$	5.15	$2.04 \cdot 10^{-15}$	14.69
PbHimda	30	0.1	"	$3.16 \cdot 10^{-10}$	9.50	$3.16 \cdot 10^{-10}$	9.50
Pb(Himda)$_2^{2-}$	30	0.1	"	$8.8 \cdot 10^{-5}$	4.17	$2.14 \cdot 10^{-14}$	13.67
ZnHimda	30	0.1	"	$2.7 \cdot 10^{-9}$	8.57	$2.7 \cdot 10^{-9}$	8.57
Zn(Himda)$_2^{2-}$	30	0.1	"	$8.0 \cdot 10^{-5}$	4.10	$2.14 \cdot 10^{-13}$	12.67

REFERENCE

1. S. CHABERECK, R. C. COURNEY and A. E. MARTELL, J. Am. Chem. Soc., 74, 5057 (1952).

Complexes with Iminodiacetic Acid

$$NH(CH_2COO)_2^{2-} \text{ (Imda}^{2-})$$

Complex ion	Temp. °C	Ionic strength	Method	h	pk	K	pK
Cdlmda	30	0.1	pH-potent.	$4.5 \cdot 10^{-6}$	5.35	$4.5 \cdot 10^{-6}$	5.35
Cd (Imda)$_2^{2-}$	30	0.1	"	$6.6 \cdot 10^{-5}$	4.18	$2.95 \cdot 10^{-10}$	9.53
Colmda	30	0.1	"	$1.12 \cdot 10^{-7}$	6.95	$1.12 \cdot 10^{-7}$	6.95
Co (Imda)$_2^{2-}$	30	0.1	"	$4.6 \cdot 10^{-6}$	5.34	$5.1 \cdot 10^{-13}$	12.29
Culmda	30	0.1	"	$2.8 \cdot 10^{-11}$	10.55	$2.8 \cdot 10^{-11}$	10.55
Cu (Imda)$_2^{2-}$	30	0.1	"	$2.24 \cdot 10^{-6}$	5.65	$6.3 \cdot 10^{-17}$	16.20
Mglmda	30	0.1	"	$2.5 \cdot 10^{-4}$	3.6	$2.5 \cdot 10^{-4}$	3.6
Nilmda	30	0.1	"	$6.2 \cdot 10^{-9}$	8.21	$6.2 \cdot 10^{-9}$	8.21
Ni (Imda)$_2^{2-}$	30	0.1	"	$4.5 \cdot 10^{-7}$	6.35	$2.76 \cdot 10^{-15}$	14.56
Znlmda	30	0.1	"	$9.3 \cdot 10^{-8}$	7.03	$9.3 \cdot 10^{-8}$	7.03
Zn (Imda)$_2^{2-}$	30	0.1	"	$7.2 \cdot 10^{-6}$	5.14	$8.8 \cdot 10^{-13}$	12.17

REFERENCE

1. S. CHABERECK, Jr., and A. E. MARTELL, J. Am. Chem. Soc., 74, 5052 (1952).

Complexes with Iminodipropionic Acid

$$NH(C_2H_4COO)_2^{2-} \text{ (Imdp}^{2-})$$

Complex ion	Temp. °C	Ionic strength	Method	h	pk	K	pK
Cdlmdp	30	0.1	pH-potent	$3.10 \cdot 10^{-4}$	3.51	$3.10 \cdot 10^{-4}$	3.51
Colmdp	30	0.1	"	$1.20 \cdot 10^{-5}$	4.92	$1.20 \cdot 10^{-5}$	4.92
Co (Imdp)$_2^{2-}$	30	0.1	"	$5.50 \cdot 10^{-4}$	3.26	$6.60 \cdot 10^{-9}$	8.18
Culmdp	30	0.1	"	$4.40 \cdot 10^{-10}$	9.36	$4.40 \cdot 10^{-10}$	9.36
Cu (Imdp)$_2^{2-}$	30	0.1	"	$2.10 \cdot 10^{-4}$	3.68	$9.10 \cdot 10^{-14}$	3.04
Nilmdp	30	0.1	"	$7.20 \cdot 10^{-7}$	6.14	$7.20 \cdot 10^{-7}$	6.14
Ni (Imdp)$_2^{2-}$	30	0.1	"	$1.70 \cdot 10^{-4}$	3.77	$1.23 \cdot 10^{-10}$	9.91
Znlmdp	30	0.1	"	$1.12 \cdot 10^{-5}$	4.95	$1.12 \cdot 10^{-5}$	4.95

REFERENCE

1. S. CHABERECK, Jr., and A. E. MARTELL, J. Am. Chem. Soc., 74, 5052 (1952).

Complexes with Iminopropionicacetic Acid

$$HN \diagdown \begin{array}{c} CH_2COO^- \\ C_2H_4COO^- \end{array}$$

Complex ion	Temp. °C	Ionic strength	Method	h	pk	K	pK
Cd Impa	30	0.1	pH-potent	$3.0 \cdot 10^{-5}$	4.52	$3.0 \cdot 10^{-5}$	4.52
Cd $(Impa)_2^{2-}$	30	0.1	"	$6.9 \cdot 10^{-4}$	3.16	$2.1 \cdot 10^{-8}$	7.68
Co Impa	30	0.1	"	$6.8 \cdot 10^{-7}$	6.17	$6.8 \cdot 10^{-7}$	6.17
Co $(Impa)_2^{2-}$	30	0.1	"	$5.1,10^{-5}$	4.29	$3.5 \cdot 10^{-11}$	10.46
Cu Impa	30	0.1	"	$3.5 \cdot 10^{-11}$	10.45	$3.5 \cdot 10^{-11}$	10.45
Cu $(Impa)_2^{2-}$	30	0.1	"	$3.5 \cdot 10^{-5}$	4.45	$1.26 \cdot 10^{-15}$	14.90
Ni Impa	30	0.1	"	$4.5 \cdot 10^{-8}$	7.35	$4.5 \cdot 10^{-8}$	7.35
Ni $(Impa)_2^{2-}$	30	0.1	"	$5.9 \cdot 0^{-6}$	5.23	$2.6 \cdot 10^{-13}$	12.58
Zn Impa	30	0.1	"	$6.8 \cdot 10^{-7}$	6.17	$6.8 \cdot 10^{-7}$	6.17
Zn $(Impa)_2^{2-}$	30	0.1	"	$4.9 \cdot 10^{-5}$	4.31	$3.3 \cdot 10^{-11}$	10.48

REFERENCE

1. S. CHABERECK, Jr., and A. E. MARTELL, *J. Am. Chem. Soc.*, **74**, 6021 (1952).

Complexes with Nitrilodiaceticpropionic Acid

$$N \diagdown \begin{array}{c} CH_2CH_2COO^- \\ — CH_2COO^- \quad (Ndap^{3-}) \\ CH_2COO^- \end{array}$$

Complex ion	Temp. °C	Ionic strength	Method	h	pk	K	pK
CdNdap⁻	30	0.1	pH-potent.	$3.2 \cdot 10^{-8}$	7.5	$3.2 \cdot 10^{-8}$	7.5
CoNdap⁻	30	0.	"	$8.0 \cdot 10^{-11}$	10.1	$8.0 \cdot 10^{-11}$	10.1
CuNdap⁻	30	0.1	"	$1.26 \cdot 10^{-12}$	11.9	$1.26 \cdot 10^{-12}$	11.9
MgNdap⁻	30	0.1	"	$6.3 \cdot 10^{-6}$	5.2	$6.3 \cdot 10^{-6}$	5.2
NiNdap⁻	30	0.1	"	$4.0 \cdot 10^{-12}$	11.4	$4.0 \cdot 10^{-12}$	11.4
ZnNdap⁻	30	0.1	"	$8.0 \cdot 10^{-11}$	10.1	$8.0 \cdot 10^{-11}$	10.1

REFERENCE

1. S. CHABERECK, Jr., and A. E. MARTELL, *J. Am. Chem. Soc.*, **75**, 2888 (1953).

Complexes with Nitrilodipropionicacetic Acid

$$N \begin{array}{l} \diagup CH_2CH_2COO^- \\ - CH_2COO^- \quad (Ndpa^{3-}) \\ \diagdown CH_2CH_2COO^- \end{array}$$

Complex ion	Temp. °C	Ionic strength	Method	k	pk	K	pK
CdNdpa⁻	30	0.1	pH-potent.	$2.5 \cdot 10^{-6}$	5.6	$2.5 \cdot 10^{-6}$	5.6
CoNdpa⁻	30	0.1	"	$1.26 \cdot 10^{-8}$	7.9	$1.26 \cdot 10^{-8}$	7.9
CuNdpa⁻	30	0.1	"	$1.26 \cdot 10^{-12}$	11.9	$1.26 \cdot 10^{-12}$	11.9
MgNdpa⁻	30	0.1	"	$2.5 \cdot 10^{-4}$	3.6	$2.5 \cdot 10^{-4}$	3.6
NiNdpa⁻	30	0.1	"	$8.0 \cdot 10^{-10}$	9.1	$8.0 \cdot 10^{-10}$	9.1
ZnNdpa⁻	30	0.1	"	$1.0 \cdot 10^{-8}$	8.0	$1.0 \cdot 10^{-8}$	8.0

REFERENCE

1. S. CHABERECK, Jr., and A. E. MARTELL, J. Am. Chem. Soc., 75, 2888 (1953).

Complexes with Nitrilotripropionic Acid

$$N \begin{array}{l} \diagup CH_2CH_2COO^- \\ - CH_2CH_2COO^- \quad (Ntp^{3-}) \\ \diagdown CH_2CH_2COO^- \end{array}$$

Complex ion	Temp. °C	Ionic strength	Method	k	pk	K	pK
CdNtp⁻	30	0.1	pH-potent.	$4.0 \cdot 10^{-4}$	3.4	$4.0 \cdot 10^{-4}$	3.4
CoNtp⁻	30	0.1	"	$1.6 \cdot 10^{-5}$	4.8	$1.6 \cdot 10^{-5}$	4.8
CuNtp⁻	30	0.1	"	$8.0 \cdot 10^{-10}$	9.1	$8.0 \cdot 10^{-10}$	9.1
NiNtp⁻	30	0.1	"	$1.6 \cdot 10^{-6}$	5.8	$1.6 \cdot 10^{-6}$	5.8
ZnNtp⁻	30	0.1		$5.0 \cdot 10^{-6}$	5.3	$5.0 \cdot 10^{-6}$	5.3

REFERENCE

1. S. CHABERECK, Jr., and A. E. MARTELL, J. Am. Chem. Soc., 75, 2888 (1953).

Complexes with Nitrilotriacetic Acid

$$N{-}CH_2COO^- \begin{smallmatrix} \diagup CH_2COO^- \\ \diagdown CH_2COO^- \end{smallmatrix} \quad (Nta^{3-})$$

Complex ion	Temp. °C	ionic strength	Method	k	pk	K	pK	References principal	References supplementary
BaNta⁻	20	0	pH-potent.	$3.9 \cdot 10^{-7}$	6.41	$3.9 \cdot 10^{-7}$	6.41	[1]	[5]
CaNta⁻	20	0	"	$6.6 \cdot 10^{-9}$	8.18	$6.6 \cdot 10^{-9}$	8.18	[1]	[3,5]
Ca(Nta)₂⁴⁻	20	0	"	$3.7 \cdot 10^{-4}$	3.43	$2.44 \cdot 10^{-12}$	11.61	[1]	[3]
CdNta⁻	20	0.1	"	$2.9 \cdot 10^{-10}$	9.54	$2.9 \cdot 10^{-10}$	9.54	[2]	[2]
Cd(Nta)₂⁴⁻	—	0.001	"	$2.0 \cdot 10^{-6}$	5.7	—	—	[3]	—
CeNta	—	0.001	"	$8.0 \cdot 10^{-9}$	8.1	$8.0 \cdot 10^{-9}$	8.1	[3]	—
CoNta⁻	20	0.1	"	$2.46 \cdot 10^{-11}$	10.61	$2.46 \cdot 10^{-11}$	10.61	[2]	—
Co(Nta)₂⁴⁻	—	0.001	"	$1.26 \cdot 10^{-4}$	3.9	—	—	[3]	—
CuNta⁻	20	0.1	"	$2.1 \cdot 10^{-13}$	12.68	$2.1 \cdot 10^{-13}$	12.68	[2]	—
FeNta⁻	20	0.1	"	$1.45 \cdot 10^{-9}$	8.84	$1.45 \cdot 10^{-9}$	8.84	[2]	—
Fe(Nta)₂³⁻	20	0.1	"	$6.3 \cdot 10^{-9}$	8.2	—	—	[4]	—
LaNta	20	0.1	"	$4.25 \cdot 10^{-11}$	10.37	$4.25 \cdot 10^{-11}$	10.37	[2]	—
La(Nta)₂³⁻	—	0.001	"	$4.0 \cdot 10^{-8}$	7.4	—	—	[3]	—
LiNta²⁻	20	0	"	$5.25 \cdot 10^{-4}$	3.28	$5.25 \cdot 10^{-4}$	3.28	[1]	—
MgNta⁻	20	0.1	"	$1 \cdot 10^{-7}$	7.0	$1 \cdot 10^{-7}$	7.0	[3]	[5]
Mg(Nta)₂⁴⁻	20	0.1	"	$6.3 \cdot 10^{-4}$	3.2	$6.3 \cdot 10^{-11}$	10.2	[3]	—

Complexes with Nitrilotriacetic Acid (contd.)

Complex ion	Temp. °C	Ionic strength	Method	k	pk	K	pK	References principal	References supplementary
MnNta^{4-}	20	0.1	pH-potent.	$5.64 \cdot 10^{-8}$	7.44	$5.64 \cdot 10^{-8}$	7.44	[2]	—
Mn(Nta)$_2^{4-}$	—	0.001	.	$2.0 \cdot 10^{-4}$	3.7	—	—	[3]	—
NaNta^{2-}	20	0	.	$7.1 \cdot 10^{-3}$	2.15	$7.1 \cdot 10^{-3}$	2.15	[1]	—
NiNta^{-}	20	0.1	.	$5.5 \cdot 10^{-12}$	11.26	$5.5 \cdot 10^{-12}$	11.26	[2]	—
Ni(Nta)$_2^{4-}$	—	0.001	.	$2.0 \cdot 10^{-5}$	4.7	—	—	[3]	—
PbNta^{-}	20	0.1	.	$1.6 \cdot 10^{-12}$	11.8	$1.6 \cdot 10^{-12}$	11.8	[3]	—
SrNta^{-}	20	0	.	$1.86 \cdot 10^{-7}$	6.73	$1.86 \cdot 10^{-7}$	6.73	[1]	[5]
ZnNta^{-}	20	0.1	.	$3.55 \cdot 10^{-11}$	10.45	$3.55 \cdot 10^{-11}$	10.45	[2]	[2]
Zn(Nta)$_2^{4-}$	—	0.001	.	$1 \cdot 10^{-3}$	3.0	—	—	[3]	—

REFERENCES

1. G. SCHWARZENBACH, E. KAMPITSCH and R. STEINER, Helv. chim. Acta, 28, 828 (1945).

2. G. SCHWARZENBACH and E. FREITAG, ibid., 34, 1492 (1951).

3. G. SCHWARZENBACH and W. BIEDERMAN, ibid., 31, 331 (1948).

4. J. HELLER and G. SCHWARZENBACH, ibid., 34, 1876 (1951).

5. G. SCHWARZENBACH, H. ACKERMANN and P. RUCKSTUHL, ibid., 32, 1175 (1949).

Complexes with Trimethylenediaminetetraacetic Acid

$$\begin{matrix} OOCH_2C \\ OOCH_2C \end{matrix}\!>\!N(CH_2)_3-N\!<\!\begin{matrix} CH_2COO^- \\ CH_2COO^- \end{matrix} \quad (Tmta^{4-})$$

Complex ion	Temp. °C	Ionic strength	Method	k	pk	K	pK
BaTmta²⁻	20	0.1	pH-potent.	$5.75 \cdot 10^{-6}$	4.24	$5.75 \cdot 10^{-5}$	4.24
CaTmta²⁻	20	0.1	"	$7.60 \cdot 10^{-8}$	7.12	$7.60 \cdot 10^{-8}$	7.12
MgTmta²⁻	20	0.1	"	$9.50 \cdot 10^{-7}$	6.02	$9.50 \cdot 10^{-7}$	6.02
SrTmta²⁻	20	0.1	"	$6.60 \cdot 10^{-6}$	5.18	$6.60 \cdot 10^{-6}$	5.18

REFERENCE

1. G. SCHWARZENBACH and H. ACKERMANN, Helv. chim. Acta, 31, 1029 (1948).

4. Complexes with Diketones and Aldehydes

Complexes with Acetylacetone $CH_3CO^- = CHCOCH_2$ $(Acac^-)$

Complex ion	Temp. °C	Ionic strength	Method	k	pk	K	pK	References principal	References supplementary
AlAcac²⁺	30	0	pH-potent.	$2.5 \cdot 10^{-9}$	8.6	$2.5 \cdot 10^{-9}$	8.6	[1]	[1]
Al(Acac)₂⁺	30	0	"	$1.26 \cdot 10^{-8}$	7.9	$3.14 \cdot 10^{-17}$	16.5	[1]	[1]
Al(Acac)₃	30	0	"	$1.6 \cdot 10^{-6}$	5.8	$5.0 \cdot 10^{-23}$	22.3	[1]	[1]
BeAcac⁺	20	0	"	$1.32 \cdot 10^{-8}$	7.88	$1.32 \cdot 10^{-8}$	7.88	[2]	[2]
Be(Acac)₂	20	0	"	$1.78 \cdot 10^{-7}$	6.75	$2.35 \cdot 10^{-15}$	14.63	[2]	[2]
CdAcac⁺	20	0	"	$1.44 \cdot 10^{-4}$	3.84	$1.44 \cdot 10^{-4}$	3.84	[2]	[2]
Cd(Acac)₂	20	0	"	$1.32 \cdot 10^{-3}$	2.83	$1.9 \cdot 10^{-7}$	6.72	[2]	[2]
CeAcac²⁺	20	0	"	$5.0 \cdot 10^{-6}$	5.30	$5.0 \cdot 10^{-6}$	5.30	[2]	[2]
Ce(Acac)₂⁺	20	0	"	$1.07 \cdot 10^{-4}$	3.97	$5.35 \cdot 10^{-10}$	9.27	[2]	[2]
Ce(Acac)₃	20	0	"	$4.16 \cdot 10^{-4}$	3.38	$2.23 \cdot 10^{-13}$	12.65	[2]	[2]
CoAcac⁺	20	0	"	$4.0 \cdot 10^{-6}$	5.40	$4.0 \cdot 10^{-6}$	5.40	[2]	[2]
Co(Acac)₂	20	0	"	$6.75 \cdot 10^{-5}$	4.17	$2.7 \cdot 10^{-10}$	9.57	[2]	[2]
CuAcac⁺	20	0	"	$4.9 \cdot 10^{-9}$	8.31	$4.9 \cdot 10^{-9}$	8.31	[2]	[2]
Cu(Acac)₂	20	0	"	$1.41 \cdot 10^{-7}$	6.85	$6.9 \cdot 10^{-16}$	15.16	[2]	[2]
EuAcac²⁺	30	0	"	$1 \cdot 10^{-6}$	6.0	$1 \cdot 10^{-6}$	6.0	[1]	[1]
Eu(Acac)₂⁺	30	0	"	$3.16 \cdot 10^{-5}$	4.5	$3.16 \cdot 10^{-11}$	10.5	[1]	[1]
Eu(Acac)₃	30	0	"	$3.16 \cdot 10^{-4}$	3.5	$1 \cdot 10^{-14}$	14.0	[1]	[1]
FeAcac²⁺	30	0	"	$1.6 \cdot 10^{-10}$	9.8	$1.6 \cdot 10^{-10}$	9.8	[1]	[1,3]
Fe(Acac)₂⁺	30	0	"	$1 \cdot 10^{-9}$	9.0	$1.6 \cdot 10^{-19}$	18.8	[1]	[1,3]
Fe(Acac)₃	30	0	"	$4.0 \cdot 10^{-8}$	7.4	$6.4 \cdot 10^{-27}$	26.2	[1]	[1,3]

Complexes with Acetylacetone (contd.)

Complex ion	Temp °C	Ionic strength	Method	k	pk	K	pK	References principal	supple-mentary
GaAcac²⁺	30	0	pH-potent.	$4.0 \cdot 10^{-10}$	9.4	$4.0 \cdot 10^{-10}$	9.4	[1]	[1]
Ga(Acac)₂⁺	30	0	"	$5.0 \cdot 10^{-9}$	8.3	$2.0 \cdot 10^{-18}$	17.7	[1]	[1]
Ga(Acac)₃	30	0	"	$1.26 \cdot 10^{-6}$	5.9	$2.5 \cdot 10^{-24}$	23.6	[1]	[1]
InAcac³⁺	30	0	"	$1 \cdot 10^{-8}$	8.0	$1 \cdot 10^{-8}$	8.0	[1]	[1]
In(Acac)₂⁺	30	0	"	$8 \cdot 10^{-8}$	7.1	$8 \cdot 10^{-16}$	15.1	[1]	[1]
LaAcac²⁺	30	0	"	$7.95 \cdot 10^{-6}$	5.1	$7.95 \cdot 10^{-6}$	5.1	[1]	[1]
La(Acac)₂⁺	30	0	"	$1.58 \cdot 10^{-4}$	3.8	$1.25 \cdot 10^{-9}$	8.9	[1]	[1]
La(Acac)₃	30	0	"	$1.26 \cdot 10^{-3}$	2.9	$1.6 \cdot 10^{-12}$	11.8	[1]	[1]
MgAcac⁺	20	0	"	$2.14 \cdot 10^{-4}$	3.67	$2.14 \cdot 10^{-4}$	3.67	[2]	[2]
Mg(Acac)₂	20	0	"	$1.95 \cdot 10^{-3}$	2.71	$2.16 \cdot 10^{-7}$	6.38	[2]	[2]
MnAcac⁺	20	0	"	$5.75 \cdot 10^{-5}$	4.24	$5.75 \cdot 10^{-5}$	4.24	[2]	[2]
Mn(Acac)₂	20	0	"	$7.76 \cdot 10^{-4}$	3.11	$4.45 \cdot 10^{-8}$	7.35	[2]	[2]
NdAcac²⁺	30	0	"	$2.51 \cdot 10^{-6}$	5.6	$2.51 \cdot 10^{-6}$	5.6	[1]	[1]
Nd(Acac)₂⁺	30	0	"	$5.0 \cdot 10^{-5}$	4.3	$1.25 \cdot 10^{-10}$	9.9	[1]	[1]
Nd(Acac)₃	30	0	"	$6.3 \cdot 10^{-4}$	3.2	$7.9 \cdot 10^{-14}$	13.1	[1]	[1]
NiAcac⁺	20	0	"	$8.7 \cdot 10^{-7}$	6.06	$8.7 \cdot 10^{-7}$	6.06	[2]	[2]
Ni(Acac)₂	20	0	"	$1.95 \cdot 10^{-5}$	4.71	$1.7 \cdot 10^{-11}$	10.77	[2]	[2]
Ni(Acac)₃⁻	20	0	"	$4.8 \cdot 10^{-3}$	2.32	$8.15 \cdot 10^{-14}$	13.09	[2]	[2]
ScAcac²⁺	30	0	"	$1 \cdot 10^{-8}$	8.0	$1 \cdot 10^{-8}$	8.0	[1]	[1]
Sc(Acac)₂⁺	30	0	"	$6.3 \cdot 10^{-8}$	7.2	$6.3 \cdot 10^{-16}$	15.2	[1]	[1]
SmAcac²⁺	30	0	"	$1.26 \cdot 10^{-6}$	5.9	$1.26 \cdot 10^{-6}$	5.9	[1]	[1]
Sm(Acac)₂⁺	30	0	"	$3.16 \cdot 10^{-5}$	4.5	$4.0 \cdot 10^{-11}$	10.4	[1]	[1]
Sm(Acac)₃	30	0	"	$6.3 \cdot 10^{-4}$	3.2	$2.5 \cdot 10^{-14}$	13.6	[1]	[1]
ThAcac³⁺	30	0	"	$1.6 \cdot 10^{-9}$	8.8	$1.6 \cdot 10^{-9}$	8.8	[1]	[1]

Complexes with Acetylacetone (contd.)

Complex ion	Temp. °C	Ionic strength	Method	k	pk	K	pK	References principal	supplementary
Th(Acac)₂²⁺	30	0	pH-potent	$4.0 \cdot 10^{-8}$	7.4	$6.4 \cdot 10^{-17}$	16.2	[1]	[1]
Th(Acac)₃⁺	30	0	"	$5.0 \cdot 10^{-7}$	6.3	$3.2 \cdot 10^{-23}$	22.5	[1]	[1]
Th(Acac)₄	30	0	"	$6.3 \cdot 10^{-5}$	4.2	$2.0 \cdot 10^{-27}$	26.7	[1]	[1]
UO₂Acac⁺	20	0	"	$2.19 \cdot 10^{-8}$	7.66	$2.19 \cdot 10^{-8}$	7.66	[2]	[2]
UO₂(Acac)₂	20	0	"	$3.24 \cdot 10^{-7}$	6.49	$7.1 \cdot 10^{-15}$	14.15	[2]	[2]
YAcac²⁺	30	0	"	$4.0 \cdot 10^{-7}$	6.4	$4.0 \cdot 10^{-7}$	6.4	[1]	[1]
Y(Acac)₂⁺	30	0	"	$2.0 \cdot 10^{-5}$	4.7	$8 \cdot 10^{-12}$	11.1	[1]	[1]
Y(Acac)₃	30	0	"	$1.6 \cdot 10^{-3}$	2.8	$1.26 \cdot 10^{-14}$	13.9	[1]	[1]
ZnAcac⁺	20	0	"	$8.5 \cdot 10^{-6}$	5.07	$8.5 \cdot 10^{-6}$	5.07	[2]	[2]
Zn(Acac)₂	20	0	"	$1.12 \cdot 10^{-4}$	3.95	$9.5 \cdot 10^{-10}$	9.02	[2]	[2]

REFERENCES

1. R. M. IZATT, et al., J. Phys. Chem., 59, 170 (1955).

2. R. M. IZATT, W. C. FERNELIUS and B. P. BLOCK, ibid., 59, 235 (1955).

3. J. BADOZ-LAMBLING, Ann. chimie., 8, No. 12, 586 (1953); cited in Ref. zh. khim., 13790 (1955).

Complexes with β-Methyltropolone*

$(βMet^-)$

Complex ion	Temp. °C	Ionic strength	Method	k	pk	K	pK
BeβMet⁺	30	—	pH-potent.	$4.0 \cdot 10^{-10}$	9.4	$4.0 \cdot 10^{-10}$	9.4
Be (βMet)₂	30	—	.	$2.0 \cdot 10^{-8}$	7.7	$8.0 \cdot 10^{-8}$	17.1
CaβMet⁺	30	—	.	$5.0 \cdot 0^{-4}$	5.3	$5.0 \cdot 10^{-8}$	5.3
Ca (βMet)₂	30	—	.	$2.5 \cdot 0^{-4}$	3.6	$1.3 \cdot 10^{-8}$	8.9
CoβMet⁺	30	—	.	$1.3 \cdot 10^{-8}$	7.9	$1.3 \cdot 10^{-8}$	7.9
Co (βMet)₂	30	—	.	$6.3 \cdot 10^{-7}$	6.2	$8.0 \cdot 10^{-15}$	14.1
Mgβ Met⁺	30	—	.	$1.0 \cdot 10^{-6}$	6.0	$1.0 \cdot 10^{-6}$	6.0
Mg (βMet)₂	30	—	.	$2.5 \cdot 10^{-5}$	4.6	$2.5 \cdot 10^{-11}$	10.6
Mg (βMet)₃⁻	30	—	.	$8.0 \cdot 10^{-1}$	3.1	$2.0 \cdot 10^{-14}$	13.7
Ni₃Met⁺	30	—	.	$4.0 \cdot 10^{-9}$	8.4	$4.0 \cdot 10^{-9}$	8.4
Ni (βMet)₂	30	—	.	$2.5 \cdot 10^{-7}$	6.6	$1.0 \cdot 10^{-15}$	15.0
Ni (βMet)₃⁻	30	—	.	$8.0 \cdot 10^{-5}$	4.1	$8.0 \cdot 10^{-20}$	19.1
PbβMet⁺	30	—	.	$2.5 \cdot 10^{-10}$	9.6	$2.5 \cdot 10^{-10}$	9.6
Pb (βMet)₂	30	—	.	$2.5 \cdot 10^{-7}$	6.6	$6.3 \cdot 10^{-17}$	16.2
Zn₃Met⁺	30	—	.	$4.0 \cdot 10^{-9}$	8.4	$4.0 \cdot 10^{-9}$	8.4
Zn (βMet)₂	30	—	.	$1.6 \cdot 10^{-7}$	6.8	$6.3 \cdot 10^{-16}$	15.2
Zn (βMet)₃⁻	30	—	.	$2.5 \cdot 10^{-4}$	3.6	$1.6 \cdot 10^{-19}$	18.8

* In 50% dioxan-water.

REFERENCE

1. B. E. BRYANT and W. C. FERNELIUS, J. Am. Chem. Soc., 76, 1696 (1954).

Complexes with Sulphosalicylaldehyde

$$^-O_3S-\!\!\left\langle\begin{array}{c}\rule{0pt}{8pt}\end{array}\right\rangle\!\!-O^-\ (SSald^{2-})$$
COH

Complex ion	Temp. °C	Ionic strength	Method	k	pk	K	pK
CoSSald	25	—	pH-potent.	$3.8 \cdot 10^{-4}$	3.42	$3.8 \cdot 10^{-4}$	3.42
CuSSald	25	—	"	$4.5 \cdot 10^{-6}$	5.35	$4.5 \cdot 10^{-6}$	5.35
Cu $(SSald)_2^{2-}$	25	—	"	$1.2 \cdot 10^{-4}$	3.92	$5.4 \cdot 10^{-10}$	9.27
NiSSald	25	—	"	$1.62 \cdot 10^{-4}$	3.79	$1.62 \cdot 10^{-4}$	3.79
Ni $(SSald)_2^{2-}$	25	—	"	$1.7 \cdot 10^{-3}$	2.77	$2.76 \cdot 10^{-7}$	6.56
ZnSSald	25	—	"	$1.0 \cdot 10^{-3}$	3.0	$1.0 \cdot 10^{-3}$	3.0

REFERENCE

1. M. CALVIN and N. C. MELCHIOR, J. Am. Chem. Soc., 70, 3270 (1948).

Complexes with Thenoyltrifluoroacetone

$$\left[\!\!\begin{array}{c}\rule{0pt}{8pt}\end{array}\!\!\right]_S\!\!\diagup COCHC\bar{O}CF_3(Tf\ )$$

Complex ion	Temp. °C	Ionic strength	Method	k	pk	K	pK
CuTf⁺	20	0.1	pH-potent.	$2.8 \cdot 10^{-7}$	6.55	$2.8 \cdot 10^{-7}$	6.55
Cu $(Tf)_2$	—	—	"		—	$1.0 \cdot 10^{-13}$	13.0
FeTf²⁺	25	0.1	Spectr.	$1.26 \cdot 10^{-7}$	6.9	$1.26 \cdot 10^{-7}$	6.9
Ni(Tf)₂	—	—	pH-potent.		—	$1.0 \cdot 10^{-10}$	10.00
PuTf³⁺	25	0.2	Spectr.	$1.0 \cdot 10^{-8}$	8.0	$1.0 \cdot 10^{-8}$	8.0
ThTf³⁺	25	0.11	"	$4.0 \cdot 10^{-8}$	7.4	$4.0 \cdot 10^{-8}$	7.4
UTf³⁺	25	0.1	"	$6.3 \cdot 10^{-8}$	7.2	$6.3 \cdot 10^{-8}$	7.2

REFERENCE

1. A. E. MARTELL and M. CALVIN, Chemistry of the Metal Chelate Compounds, New York (1953).

Complexes with Tropolone*

O=⬡—O⁻ (Trop⁻)

Complex ion	Temp. °C	Ionic strength	Method	k	pk	K	pK
BeTrop⁺	30	—	pH-potent.	$4.0 \cdot 10^{-9}$	8.4	$4.0 \cdot 10^{-9}$	8.4
Be(Trop)₂	30	—	"	$1.0 \cdot 10^{-7}$	7.0	$4.0 \cdot 10^{-16}$	15.4
CaTrop⁺	30	—	"	$1.6 \cdot 10^{-5}$	4.8	$1.6 \cdot 10^{-5}$	4.8
Ca(Trop)₂	30	—	"	$6.3 \cdot 10^{-4}$	3.2	$1.0 \cdot 10^{-8}$	8.0
CoTrop⁺	30	—	"	$1.0 \cdot 10^{-7}$	7.0	$1.0 \cdot 10^{-7}$	7.0
Co(Trop)₂	30	—	"	$1.26 \cdot 10^{-6}$	5.9	$1.26 \cdot 10^{-13}$	12.9
Co(Trop)₃⁻	30	—	"	$1.6 \cdot 10^{-4}$	3.8	$2.0 \cdot 10^{-17}$	16.7
Cu(Trop)₃	30	—	"	$1.26 \cdot 10^{-8}$	7.9	—	—
MgTrop⁺	30	—	"	$3.2 \cdot 10^{-6}$	5.5	$3.2 \cdot 10^{-6}$	5.5
Mg(Trop)₂	30	—	"	$4.0 \cdot 10^{-5}$	4.4	$1.3 \cdot 10^{-10}$	9.9
NiTrop⁺	30	—	"	$2.0 \cdot 10^{-8}$	7.7	$2.0 \cdot 10^{-8}$	7.7
Ni(Trop)₂	30	—	"	$8.0 \cdot 10^{-7}$	6.1	$1.6 \cdot 10^{-14}$	13.8
Ni(Trop)₃⁻	30	—	"	$1.0 \cdot 10^{-4}$	4.0	$1.6 \cdot 10^{-18}$	17.8
PbTrop⁺	30	—	"	$1.0 \cdot 10^{-8}$	8.0	$1.0 \cdot 10^{-8}$	8.0
Pb(Trop)₂	30	—	"	$1.0 \cdot 10^{-6}$	6.0	$1.0 \cdot 10^{-14}$	14.0
ZnTrop⁺	30	—	"	$3.2 \cdot 10^{-8}$	7.5	$3.2 \cdot 10^{-8}$	7.5
Zn(Trop)₂	30	—	"	$3.2 \cdot 10^{-7}$	6.5	$1.0 \cdot 10^{-14}$	14.0
Zn(Trop)₃⁻	30	—	"	$3.2 \cdot 10^{-4}$	3.5	$3.2 \cdot 10^{-18}$	17.5

* In 50% dioxan-water.

REFERENCE

1. B. E. BRYANT, W. C. FERNELIUS and B. E. DOUGLAS, J. Am. Chem. Soc., 75, 3784 (1953).

Complexes with Salicylaldehyde

—CHO (Sald⁻), O⁻

Complex ion	Temp °C	Ionic strength	Method	k	pH	K	pK	References principal	References supplementary
Cd (Sald)₂	20	0.1	pH-potent	—	—	$1.6 \cdot 10^{-8}$	7.8	[1]	
CoSald⁺	20	0.1	.	$2.14 \cdot 10^{-5}$	4.67	$2.14 \cdot 10^{-9}$	4.67	[1]	
Có (Sald)₂	20	0.1	.	$2.34 \cdot 10^{-4}$	3.63	$5.0 \cdot 10^{-9}$	8.3	[1]	
CuSald⁺	20	0.1	.	$4.0 \cdot 10^{-8}$	7.40	$4.0 \cdot 10^{-8}$	7.40	[1]	
Cu (Sald)₂	20	0.1	.	$1.26 \cdot 10^{-6}$	5.90	$5.0 \cdot 10^{-14}$	13.3	[1]	[2]
FeSald⁺	20	0.1	.	$6.0 \cdot 10^{-5}$	4.22	$6.0 \cdot 10^{-5}$	4.22	[1]	
Fe (Sald)₂	20	0.1	.	$4.17 \cdot 10^{-4}$	3.38	$2.5 \cdot 10^{-8}$	7.6	[1]	
MnSald⁺	20	0.1	.	$1.86 \cdot 10^{-4}$	3.73	$1.86 \cdot 10^{-4}$	3.73	[1]	
Mn (Sald)₂	20	0.1	.	$8.5 \cdot 10^{-4}$	3.07	$1.6 \cdot 10^{-7}$	6.8	[1]	
NiSald⁺	20	0.1	.	$6.0 \cdot 10^{-5}$	5.22	$6.0 \cdot 10^{-6}$	5.22	[1]	
Ni (Sald)₂	20	0.1	.	$1.05 \cdot 10^{-4}$	3.98	$6.3 \cdot 10^{-10}$	9.2	[1]	
Pb (Sald)₂	20	0.1	.	—	—	$8.0 \cdot 10^{-10}$	9.1	[1]	
ZnSald⁺	20	0.1	.	$3.16 \cdot 10^{-5}$	4.50	$3.16 \cdot 10^{-5}$	4.50	[1]	
Zn (Sald)₂	20	0.1	.	$2.5 \cdot 10^{-4}$	3.80	$7.9 \cdot 10^{-9}$	8.1	[1]	

REFERENCES

1. D. C. MELLOR and L. MALLEY, Nature, 159, 370 (1947).

2. M. CALVIN and K. W. WILSON, J. Am. Chem. Soc., 67, 2003 (1945).

Complexes with α-Isopropyltropolone*   (αIst⁻)

$$\text{O} \quad \text{O}^-$$
$$\text{CH(CH}_3)_2$$

Complex ion	Temp. °C	Ionic strength	Method	k	pk	K	pK
BeαIst⁺	30	—	pH-potent.	$2.0\cdot10^{-11}$	10.7	$2.0\cdot10^{-11}$	10.7
Be(αIst)₂	30	—		$8.0\cdot10^{-10}$	9.1	$1.6\cdot10^{-20}$	19.8
CaαIst⁺	30	—		$6.3\cdot10^{-6}$	5.2	$6.3\cdot10^{-6}$	5.2
Ca(αIst)₂	30	—		$2.5\cdot10^{-4}$	3.6	$1.6\cdot10^{-9}$	8.8
CoαIst⁺	30	—		$8.0\cdot10^{-9}$	8.1	$8.0\cdot10^{-9}$	8.1
Co(αIst)₂	30	—		$2.0\cdot10^{-7}$	6.7	$4.0\cdot10^{-14}$	13.4
MgαIst⁺	30	—		$6.3\cdot10^{-7}$	6.2	$6.3\cdot10^{-7}$	6.2
Mg(αIst)₂	30	—		$6.3\cdot10^{-6}$	5.2	$4.0\cdot10^{-12}$	11.4
NiαIst⁺	30	—		$2.5\cdot10^{-9}$	8.6	$2.5\cdot10^{-9}$	8.6
Ni(αIst)₂	30	—		$1.3\cdot10^{-7}$	6.9	$3.2\cdot10^{-16}$	15.5
Ni(αIst)₃⁻	30	—		$2.0\cdot10^{-4}$	3.7	$6.3\cdot10^{-20}$	19.2
PbαIst⁺	30	—		$3.2\cdot10^{-10}$	9.5	$3.2\cdot10^{-10}$	9.5
Pb(αIst)₂	30	—		$3.2\cdot10^{-8}$	7.5	$1.0\cdot10^{-17}$	17.0
ZnαIst⁺	30	—		$2.0\cdot10^{-9}$	8.7	$2.0\cdot10^{-9}$	8.7
Zn(αIst)₂	30	—		$3.2\cdot10^{-8}$	7.5	$6.3\cdot10^{-17}$	16.2

* In 50% dioxan-water.

REFERENCE

1. B. E. BRYANT and W. C. FERNELIUS, J. Am. Chem. Soc., 76, 1696 (1954).

Complexes with β-Isopropyltropolone*   (βIst-)

Complex ion	Temp. °C	Ionic strength	Method	k	pk	K	pK
Be₁lst⁺	30	—	pH-potent.	$8.0 \cdot 10^{-10}$	9.1	$8.0 \cdot 10^{-10}$	9.1
Be (β1st)₂	30	—	"	$3.2 \cdot 10^{-8}$	7.5	$2.5 \cdot 10^{-17}$	16.6
Ca₂lst⁺	30	—	"	$4.0 \cdot 10^{-6}$	5.4	$4.0 \cdot 10^{-6}$	5.4
Ca (β1st)₂	30	—	"	$2.5 \cdot 10^{-4}$	3.6	$1.0 \cdot 10^{-9}$	9.0
Co₂lst⁺	30	—	"	$1.3 \cdot 10^{-8}$	7.9	$1.3 \cdot 10^{-8}$	7.9
Co (β1st)₂	30	—	"	$5.0 \cdot 10^{-7}$	6.3	$6.3 \cdot 10^{-15}$	14.2
Co (β1st)₃⁻	30	—	"	$1.6 \cdot 10^{-4}$	3.8	$1.0 \cdot 10^{-18}$	18.0
Mg₂lst⁺	30	—	"	$6.3 \cdot 10^{-7}$	6.2	$6.3 \cdot 10^{-7}$	6.2
Mg (β1st)₂	30	—	"	$1.6 \cdot 10^{-5}$	4.8	$1.0 \cdot 10^{-11}$	11.0
Mg (β1st)₃⁻	30	—	"	$1.0 \cdot 10^{-8}$	3.0	$1.0 \cdot 10^{-14}$	14.0
Ni₁lst⁺	30	—	"	$3.2 \cdot 10^{-9}$	8.5	$3.2 \cdot 10^{-9}$	8.5
Ni (β1st)₂	30	—	"	$3.2 \cdot 10^{-7}$	6.5	$1.0 \cdot 10^{-16}$	15.0
Ni (β1st)₃⁻	30	—	"	$1.0 \cdot 10^{-4}$	4.0	$1.0 \cdot 10^{-19}$	19.0
Pb₂lst⁺	30	—	"	$1.0 \cdot 10^{-9}$	9.0	$1.0 \cdot 10^{-9}$	9.0
Pb (β1st)₂	30	—	"	$2.0 \cdot 10^{-6}$	6.7	$2.0 \cdot 10^{-16}$	15.7
Zn₂lst⁺	30	—	"	$2.0 \cdot 10^{-9}$	8.7	$2.0 \cdot 10^{-9}$	8.7
Zn (β1st)₂	30	—	"	$1.0 \cdot 10^{-7}$	7.0	$2.0 \cdot 10^{-16}$	15.7
Zn (β1st)₃⁻	30	—	"	$2.5 \cdot 10^{-4}$	3.6	$5.0 \cdot 10^{-20}$	19.3

* In 50% dioxan-water.

REFERENCE

1. B. E. BRYANT and W. C. FERNELIUS, J. Am. Chem. Soc., 76, 1696 (1954).

Complexes with α-Methyltropolone*

(αMet⁻)

Complex ion	Temp. °C	Ionic strength	Method	k	pk	K	pK
BeαMet⁺	30	—	pH-potent.	$5.0 \cdot 10^{-11}$	10.3	$5.0 \cdot 10^{-11}$	10.3
Be(αMet)₂	30	—	"	$1.0 \cdot 10^{-9}$	9.0	$5.0 \cdot 10^{-20}$	19.3
CaαMet⁺	30	—	"	$8.0 \cdot 10^{-6}$	5.1	$8.0 \cdot 10^{-6}$	5.1
Ca(αMet)₂	30	—	"	$4.0 \cdot 10^{-4}$	3.4	$3.2 \cdot 10^{-9}$	8.5
CoαMet⁺	30	—	"	$1.0 \cdot 10^{-8}$	8.0	$1.0 \cdot 10^{-8}$	8.0
Co(αMet)₂	30	—	"	$5.0 \cdot 10^{-7}$	6.3	$5.0 \cdot 10^{-15}$	14.3
MgαMet⁺	30	—	"	$1 \cdot 10^{-6}$	6.0	$1.0 \cdot 10^{-6}$	6.0
Mg(αMet)₂	30	—	"	$2.5 \cdot 10^{-5}$	4.6	$2.5 \cdot 10^{-11}$	10.6
Mg(αMet)₃⁻	30	—	"	$2.5 \cdot 10^{-3}$	2.6	$1.6 \cdot 10^{-13}$	12.8
NiαMet⁺⁺	30	—	"	$4.0 \cdot 10^{-9}$	8.4	$4.0 \cdot 10^{-9}$	8.4
Ni(αMet)₂	30	—	"	$2.0 \cdot 10^{-7}$	6.7	$8.0 \cdot 10^{-16}$	15.1
Ni(αMet)₃⁻	30	—	"	$2.0 \cdot 10^{-4}$	3.7	$1.6 \cdot 10^{-19}$	18.8
PbαMet⁺	30	—	"	$4.0 \cdot 10^{-10}$	9.4	$4.0 \cdot 10^{-10}$	9.4
Pb(αMet)₂	30	—	"	$2.0 \cdot 10^{-7}$	6.7	$8.0 \cdot 10^{-17}$	16.1
ZnαMet⁺	30	—	"	$2.5 \cdot 10^{-9}$	8.6	$2.5 \cdot 10^{-9}$	8.6
Zn(αMet)₂	30	—	"	$8.0 \cdot 10^{-8}$	7.1	$2.0 \cdot 10^{-16}$	15.7

* In 50% dioxan-water.

REFERENCE

1. B. E. BRYANT and W. C. FERNELIUS, J. Am. Chem. Soc., 76, 1696 (1954).

5. Complexes with other Organic Ligands

Complexes with o-Aminophenol*

(Af)

Complex ion	Temp. °C	Ionic strength	Method	k	pk	K	pK
CoAf⁺	25	0.01	pH-potent.	$1.38 \cdot 10^{-6}$	5.86	$1.38 \cdot 10^{-6}$	5.86
Co(Af)₂	25	0.01	"	$2.7 \cdot 10^{-5}$	4.57	$3.70 \cdot 10^{-11}$	10.43
CuAf⁺	25	0.01	"	$5.6 \cdot 10^{-10}$	9.25	$5.6 \cdot 10^{-10}$	9.25
Cu(Af)₂	25	0.01	"	$3.4 \cdot 10^{-9}$	8.47	$1.90 \cdot 10^{-18}$	17.72
Ni Af⁺	25	0.01	"	$2.4 \cdot 10^{-7}$	6.62	$2.4 \cdot 10^{-7}$	6.62
Ni(Af)₂	25	0.01	"	$1.62 \cdot 10^{-5}$	4.79	$3.9 \cdot 10^{-12}$	11.41
PbAf⁺	25	0.01	"	$6.6 \cdot 10^{-8}$	7.18	$6.6 \cdot 10^{-8}$	7.18
Pb(Af)₂	25	0.01	"	$4.16 \cdot 10^{-6}$	5.38	$2.75 \cdot 10^{-13}$	12.56
ZnAf⁺	25	0.01	"	$4.07 \cdot 10^{-7}$	6.39	$4.07 \cdot 10^{-7}$	6.39
Zn(Af)₂	25	0.01	"	$1.7 \cdot 10^{-6}$	5.77	$6.9 \cdot 10^{-13}$	12.16

* In 50% dioxan-water.

REFERENCE

1. R. G. CHARLES and H. FREISER, J. Am. Chem. Soc., 74, 1385 (1952).

Complexes with Eriochrome Black A

(Esa³⁻)

Complex ion	Temp. °C	Ionic strength	Method	k	pk	K	pK
CaEsa⁻	Room	0.02	Spectr.	$5.6 \cdot 10^{-6}$	5.25	$5.6 \cdot 10^{-6}$	5.25
MgEsa⁻	"	0.08	"	$6.3 \cdot 10^{-8}$	7.2	$6.3 \cdot 10^{-8}$	7.2

REFERENCE

1. G. SCHWARZENBACH and W. BIEDERMAN, Helv. chim. Acta, 31, 678 (1948).

Complexes with Eriochrome Black **T**

(Est^{3-})

Complex ion	Temp. °C	Ionic strength	Method	k	pk	K	pK
CaEst⁻	Room	0.02	Spectr.	$4.0 \cdot 10^{-6}$	5.4	$4.0 \cdot 10^{-6}$	5.4
MgEst⁻	"	0.08	"	$1.0 \cdot 10^{-7}$	7.0	$1.0 \cdot 10^{-7}$	7.0

REFERENCE

1. G. SCHWARZENBACH and W. BIEDERMAN, **Helv. chim. Acta**, **31**, 678 (1948).

Complexes with Eriochrome Blue Black **B**

(Esb^{3-})

Complex ion	Temp. °C	Ionic strength	Method	k	pk	K	pK
CaEsb⁻	Room	0.02	Spectr.	$2.0 \cdot 10^{-6}$	5.7	$2.0 \cdot 10^{-6}$	5.7
MgEsb⁻	"	0.08	"	$4.0 \cdot 10^{-6}$	7.4	$4.0 \cdot 10^{-8}$	7.4

REFERENCE

1. G. SCHWARZENBACH and W. BIEDERMAN, **Helv. chim. Acta**, **31**, 678 (1948).

Complexes with Eriochrome Blue Black R

$$-O_3S-\langle\rangle N=N-\langle\rangle \quad (Ers^{3-})$$

Complex ion	Temp. °C	Ionic strength	Method	k	pk	K	pK
Ca Ers⁻	Room	0.02	Spectr.	$5.6 \cdot 10^{-6}$	5.25	$5.6 \cdot 10^{-6}$	5.25
Mg Ers⁻	"	0.08	"	$2.75 \cdot 10^{-8}$	7.56	$2.75 \cdot 10^{-8}$	7.56

REFERENCE

1. G. SCHWARZENBACH and W. BIEDERMAN, Helv. chim. Acta, 31, 678 (1948).

Complexes with 8-Hydroxy-2,4-dimethylquinazoline*

$$(DmHas^-)$$

Complex ion	Temp. °C	Ionic strength	Method	k	pk	K	pK
CuDmHas⁺	20	0.3	pH-potent.	$5.5 \cdot 10^{-11}$	10.26	$5.5 \cdot 10^{-11}$	10.26
Cu(DmHas)₂	20	0.3	"	$1.8 \cdot 10^{-10}$	9.74	$1.0 \cdot 10^{-20}$	20.00
MgDmHas⁺	20	0.3	"	$1.55 \cdot 10^{-4}$	3.81	$1.55 \cdot 10^{-4}$	3.81
Mg(DmHas)₂	20	0.3	"	$8.1 \cdot 10^{-4}$	3.09	$1.25 \cdot 10^{-7}$	6.90
NiDmHas⁺	20	0.3	"	$1.32 \cdot 10^{-8}$	7.88	$1.32 \cdot 10^{-8}$	7.88
Ni(DmHas)₂	20	0.3	"	$1.0 \cdot 10^{-7}$	7.00	$1.32 \cdot 10^{-15}$	14.88
UO₂DmHas⁺	20	0.3	"	$1.7 \cdot 10^{-9}$	8.77	$1.7 \cdot 10^{-9}$	8.77
UO₂(DmHas)	20	0.3	"	$4.7 \cdot 10^{-8}$	7.33	$8.0 \cdot 10^{-17}$	16.10
ZnDmHas⁺	20	0.3	"	$1.7 \cdot 10^{-8}$	7.77	$1.7 \cdot 10^{-8}$	7.77
Zn(DmHas)₂	20	0.3	"	$9.8 \cdot 10^{-8}$	7.81	$1.65 \cdot 10^{-15}$	14.78

* In 50% dioxan-water.

REFERENCE

1. H. IRVING and H. S. ROSSOTTI, J. Chem. Soc., 2910 (1954).

Complexes with 8-Hydroxy-4-methyl-2-phenylquinazoline*

$$CH_3$$

(MpHas⁻)

C_6H_5

O^-

Complex ion	Temp. °C	Ionic strength	Method	k	pk	K	pK
CuMpHas⁺	20	0.3	pH-potent.	$1.48 \cdot 10^{-9}$	8.83	$1.48 \cdot 10^{-9}$	8.83
Cu (MpHas)₂	20	0.3	"	$4.9 \cdot 10^{-9}$	8.31	$7.25 \cdot 10^{-18}$	17.14
UO₂MpHas⁺	20	0.3	"	$2.95 \cdot 10^{-9}$	8.53	$2.95 \cdot 10^{-9}$	8.53
UO₂(MpHas)₂	20	0.3	"	$1.41 \cdot 10^{-8}$	7.85	$4.16 \cdot 10^{-17}$	16.38
ZnMpHas⁺	20	0 3	"	$9.3 \cdot 10^{-8}$	7.03	$9.3 \cdot 10^{-8}$	7.03
Zn (MpHas)₂	20	0.3	,,	$1.35 \cdot 10^{-6}$	5.87	$1.26 \cdot 10^{-13}$	12.90

* In 50% dioxan-water.

REFERENCE

1. H. IRVING and H. S. ROSSOTTI, J. Chem. Soc., 2910 (1954).

Complexes with 8-Hydroxy-5-methylquinoline*

$$CH_3$$

(mOxin⁻)

O^- N

Complex ion	Temp. °C	Ionic strength	Method	k	pk	K	pK
Cu mOxin⁺	20	0.3	pH-potent	$2.8 \cdot 10^{-14}$	13.55	$2.8 \cdot 10^{-14}$	13.55
Cu(mOxin)₂	20	0.3	"	$4.5 \cdot 10^{-13}$	12.35	$1.26 \cdot 10^{-26}$	25.90
Mg mOxin⁺	20	0.3	"	$6.16 \cdot 10^{-6}$	5.21	$6.16 \cdot 10^{-6}$	5 21
Mg (mOxin)₂	20	0 3	"	$3.4 \cdot 10^{-5}$	4.47	$2.1 \cdot 10^{-10}$	9.68
UO₂ mOxin⁺	20	0.3	"	$5.6 \cdot 10^{-12}$	11.25	$5.6 \cdot 10^{-12}$	11.25
UO₂ (mOxin)₂	20	0.3	"	$3.0 \cdot 10^{-10}$	9.52	$1.7 \cdot 10^{-21}$	20.77
Zn mOxin⁺	20	0.3	"	$6.2 \cdot 10^{-6}$	5.21	$6.2 \cdot 10^{-6}$	5.21
Zn (mOxin)₂	20	0.3	"	$3.4 \cdot 10^{-5}$	4.47	$2.1 \cdot 10^{-10}$	9.68

* In 50% dioxan-water.

REFERENCE

1. H. IRVING and H. S. ROSSOTTI, J. Chem. Soc., 2910 (1954).

Complexes with 8-Hydroxy-2-methylquinoline*

$O^- N \; CH_3 \; (Oxd^-)$

Complex ion	Temp °C	Ionic strength	Method	k	pk	K	pK	References principal	supplementary
CeOxd³⁺	25	—	pH-potent.	$1.95 \cdot 10^{-8}$	7.71	$1.95 \cdot 10^{-8}$	7.71	[1]	—
CoOxd⁺	25	—	.	$2.10 \cdot 10^{-10}$	9.68	$2.10 \cdot 10^{-10}$	9.68	[1]	—
Co(Oxd)₂	25	—	.	$1.2 \cdot 10^{-9}$	8.92	$2.5 \cdot 10^{-19}$	18.60	[1]	[1]
CuOxd⁺	20	0.3	.	$6.0 \cdot 10^{-11}$	10.22	$6.0 \cdot 10^{-11}$	10.22	[2]	[1]
Cu(Oxd)₂	20	0.3	.	$4.8 \cdot 10^{-10}$	9.32	$2.9 \cdot 10^{-20}$	19.54	[2]	—
MgOxd⁺	20	0.3	.	$1.86 \cdot 10^{-4}$	3.73	$1.86 \cdot 10^{-4}$	3.73	[2]	—
Mg(Oxd)₂	20	0.3	.	$7.4 \cdot 10^{-4}$	3.13	$1.38 \cdot 10^{-7}$	6.86	[2]	—
MnOxd⁺	25	—	.	$1.9 \cdot 10^{-8}$	7.72	$1.9 \cdot 10^{-8}$	7.72	[1]	—
Mn(Oxd)₂	25	—	.	$1.45 \cdot 10^{-7}$	6.84	$2.76 \cdot 10^{-15}$	14.56	[1]	—
NiOxd⁺	20	0.3	.	$3.0 \cdot 10^{-9}$	8.52	$3.0 \cdot 10^{-9}$	8.52	[2]	[1]
Ni(Oxd)₂	20	0.3	.	$1.1 \cdot 10^{-8}$	7.96	$3.3 \cdot 10^{-17}$	16.48	[2]	[1]
PbOxd⁺	25	—	.	$4.5 \cdot 10^{-11}$	10.35	$4.5 \cdot 10^{-11}$	10.35	[1]	—
Pb(Oxd)₂	25	—	.	$5.6 \cdot 10^{-9}$	8.25	$2.5 \cdot 10^{-19}$	18.60	[1]	—
UO₂Oxd⁺	20	0.3	.	$4.0 \cdot 10^{-10}$	9.4	$4.0 \cdot 10^{-10}$	9.4	[2]	—
UO₂(Oxd)₂	20	0.3	.	$1.0 \cdot 10^{-8}$	8.0	$4.0 \cdot 10^{-18}$	17.4	[2]	—
ZnOxd⁺	20	0.3	.	$2.2 \cdot 10^{-9}$	8.66	$2.2 \cdot 10^{-9}$	8.66	[2]	—
Zn(Oxd)₂	20	0.3	.	$8.0 \cdot 10^{-9}$	8.10	$1.74 \cdot 10^{-17}$	16.76	[2]	[1]

* In 50% dioxan-water.

REFERENCES

1. W. D. JOHNSTON and H. FREISER, J. Am. Chem. Soc., 74, 5239 (1952).
2. H. IRVING and H. S. ROSSOTTI, J. Chem. Soc., 2910 (1954).

Complexes with 8-Hydroxy-6-methylquinoline*

(MeOxin⁻)

Complex ion	Temp. °C	Ionic strength	Method	k	pk	K	pK
MgMeOxin⁺	20	0 3	pH-potent.	$8.14 \cdot 10^{-6}$	5.09	$8.14 \cdot 10^{-6}$	5.09
Mg(MeOxin)₂	20	0.3	"	$4.90 \cdot 10^{-5}$	4.31	$4.0 \cdot 10^{-10}$	9.40
UO₂MeOxin⁺	20	0.3	"	$1.29 \cdot 10^{-11}$	10.89	$1.29 \cdot 10^{-11}$	10.89
UO₂(MeOxin)₂	20	0.3	"	$5.5 \cdot 10^{-10}$	9.26	$7.1 \cdot 10^{-21}$	20.15

* In 50% dioxan-water.

REFERENCE

1. H. IRVING and H. S. ROSSOTTI, *J. Chem. Soc.*, 2910 (1954).

Complexes with 8-Hydroxy-7-methylquinoline*

(meOxin⁻)

Complex ion	Temp. °C	Ionic strength	Method	k	pk	K	pK
MgmeOxin⁺	20	0.3	pH-potent.	$2.3 \cdot 10^{-5}$	4.64	$2.3 \cdot 10^{-5}$	4.64
Mg(meOxin)₂	20	0.3	"	$7.6 \cdot 10^{-5}$	4.12	$1.74 \cdot 10^{-9}$	8.76
UO₂meOxin⁺	20	0.3	"	$5.25 \cdot 10^{-12}$	11.28	$5.25 \cdot 10^{-12}$	11.28
UO₂(meOxin)₂	20	0.3	"	$1.66 \cdot 10^{-10}$	9.78	$8.7 \cdot 10^{-22}$	21.06
ZnmeOxin⁺	20	0.3	"	$4.9 \cdot 10^{-10}$	9.31	$4.9 \cdot 10^{-10}$	9.31
Zn(meOxin)₂	20	0.3	"	$7.8 \cdot 10^{-9}$	8.11	$3.8 \cdot 10^{-18}$	17.42

* In 50% dioxan-water.

REFERENCE

1. H. IRVING and H. S. ROSSOTTI, *J. Chem. Soc.*, 2910 (1954).

Complexes with 8-Hydroxy-4-methylcinnoline*

Complex ion	Temp. °C	Ionic strength	Method	k	pk	K	pK
MgMcin+	20	0.3	pH-potent.	$2.2 \cdot 10^{-4}$	3.66	$2.2 \cdot 10^{-4}$	3.66
Mg(Mcin)₂	20	0.3	"	$2.62 \cdot 10^{-3}$	2.58	$5.75 \cdot 10^{-7}$	6.24
NiMcin+	20	0.3	"	$3.2 \cdot 10^{-9}$	8.5	$3.2 \cdot 10^{-9}$	8.5
Ni(Mcin)₂	20	0.3	"	$6.3 \cdot 10^{-9}$	8.2	$2.0 \cdot 10^{-17}$	16.7
UO₂Mcin+	20	0.3	"	$1.0 \cdot 10^{-9}$	9.00	$1.0 \cdot 10^{-9}$	9.00
UO₂(Mcin)₂	20	0.3	"	$5.0 \cdot 10^{-8}$	7.30	$5.0 \cdot 10^{-17}$	16.30
ZnMcin+	20	0.3	"	$6.0 \cdot 10^{-8}$	7.22	$6.0 \cdot 10^{-8}$	7.22
Zn(Mcin)₂	20	0.3	"	$3.4 \cdot 10^{-7}$	6.47	$2.04 \cdot 10^{-14}$	13.69

* In 50% dioxan-water.

REFERENCE

1.   H. IRVING and H. S. ROSSOTTI, *J. Chem. Soc.*, 2910 (1954).

Complexes with 8-Hydroxyquinazoline*

Complex ion	Temp. °C	Ionic strength	Method	k	pk	K	pK
CuHas+	20	0.3	pH-potent.	$2.76 \cdot 10^{-11}$	10.56	$2.76 \cdot 10^{-11}$	10.56
Cu(Has)₂	20	0.3	"	$2.9 \cdot 10^{-10}$	9.54	$8.0 \cdot 10^{-21}$	20.10
MgHas+	20	0.3	"	$1.3 \cdot 10^{-4}$	3.89	$1.3 \cdot 10^{-4}$	3.89
Mg(Has)₂	20	0.3	"	$1.23 \cdot 10^{-3}$	2.91	$1.6 \cdot 10^{-7}$	6 80
UO₂Has+	20	0.3	"	$1.02 \cdot 10^{-9}$	8.99	$1.02 \cdot 10^{-9}$	8.99
UO₂(Has)₂	20	0.3	"	$2.0 \cdot 10^{-8}$	7.70	$2.04 \cdot 10^{-17}$	16.69
ZnHas+	20	0.3	"	$3.3 \cdot 10^{-8}$	7.48	$3.3 \cdot 10^{-8}$	7.48
Zn(Has)₂	20	0.3	"	$1.1 \cdot 10^{-7}$	6.96	$3.6 \cdot 10^{-15}$	14.44

* In 50% dioxan-water.

REFERENCE

1.   H. IRVING and H. S. ROSSOTTI, *J. Chem. Soc.*, 2910 (1954).

Complexes with 5-Hydroxyquinoxaline*

(Hox^-)

Complex ion	Temp °C	Ionic strength	Method	k	pk	K	pK
CuHox⁺	20	0 3	pH-potent.	$2.18 \cdot 10^{-10}$	9 66	$2.18 \cdot 10^{-10}$	9.66
Cu(Hox)₂	20	0.3	"	$1.45 \cdot 10^{-9}$	8.84	$3.16 \cdot 10^{-19}$	18.50
MgHox⁺	20	0.3	"	$3.63 \cdot 10^{-4}$	3 44	$3.63 \cdot 10^{-4}$	3.44
Mg(Hox)₂	20	0.3	"	$1.12 \cdot 10^{-3}$	2.95	$4.07 \cdot 10^{-7}$	6.39
NiHox⁺	20	0 3	"	$1.62 \cdot 10^{-8}$	7.79	$1.62 \cdot 10^{-8}$	7.79
Ni(Hox)₂	20	0.3	"	$9.5 \cdot 10^{-8}$	7.02	$1.55 \cdot 10^{-15}$	14.81
UO₂Hox⁺	20	0.3	"	$4.0 \cdot 10^{-9}$	8.40	$4.0 \cdot 10^{-9}$	8.40
UO₂(Hox)₂	20	0.3	"	$3.1 \cdot 10^{-8}$	7.51	$1.23 \cdot 10^{-16}$	15.91
ZnHox⁺	20	0 3	"	$8.5 \cdot 10^{-8}$	7.07	$8.5 \cdot 10^{-8}$	7.07
Zn(Hox)₂	20	0.3	"	$1.95 \cdot 10^{-6}$	5.71	$1.65 \cdot 10^{-13}$	12.78

* In 50% dioxan-water.

REFERENCE

1. H. IRVING and H. S. ROSSOTTI, J. Chem. Soc., 2910 (1954).

Complexes with 8-Hydroxycinnoline*

(Cin^-)

Complex ion	Temp. °C	Ionic strength	Method	k	pk	K	pK
CuCin⁺	20	0.3	pH-potent	$3 3 \cdot 10^{-10}$	9.48	$3.3 \cdot 10^{-10}$	9.48
Cu(Cin)₂	20	0.3	"	$2.56 \cdot 10^{-9}$	8.59	$8.5 \cdot 10^{-19}$	18.07
MgCin⁺	20	0.3	"	$9.5 \cdot 10^{-4}$	3.02	$9.5 \cdot 10^{-4}$	3.02
Mg(Cin)₂	20	0.3	"	$6.6 \cdot 10^{-3}$	2.18	$6.3 \cdot 10^{-6}$	5.20
NiCin⁺	20	0.3	"	$5.6 \cdot 10^{-9}$	8.25	$5.6 \cdot 10^{-9}$	8.25
Ni(Cin)₂	20	0.3	"	$5.9 \cdot 10^{-8}$	7.23	$3.3 \cdot 10^{-16}$	15.48
UO₂Cin⁺	20	0.3	"	$2.1 \cdot 10^{-9}$	8.68	$2.1 \cdot 10^{-9}$	8.68
UO₂(Cin)₂	20	0.3	"	$6.9 \cdot 10^{-8}$	7.16	$1.45 \cdot 10^{-16}$	15.84
ZnCin⁺	20	0.3	"	$1.17 \cdot 10^{-7}$	6.93	$1.17 \cdot 10^{-7}$	6.93
Zn(Cin)₂	20	0.3	"	$1.6 \cdot 10^{-6}$	5.80	$1.86 \cdot 10^{-13}$	12.73

* In 50% dioxan-water.

REFERENCE

1. H. IRVING and H. S. ROSSOTTI, J. Chem. Soc., 2910 (1954).

Complexes with 8-Hydroxyquinoline* $(Oxin^-)$

Complex ion	Temp. °C	Ionic strength	Method	k	pk	K	pK	References prin-cipal	supple-mentary
$Cd\ Oxin^+$	25	—	pH-potent.	$3.7\cdot10^{-10}$	9.43	$3.7\cdot10^{-10}$	9.43	[1]	—
$Cd\ (Oxin)_2$	25	—	"	$2.09\cdot10^{-8}$	7.68	$7.75\cdot10^{-18}$	17.11	[1]	—
$Ce\ Oxin^{2+}$	25	—	"	$7.1\cdot10^{-10}$	9.15	$7.1\cdot10^{-10}$	9.15	[1]	—
$Ce\ (Oxin)_2^+$	25	—	"	$1.05\cdot10^{-8}$	7.98	$7.45\cdot10^{-18}$	17.13	[1]	—
$Co\ Oxin^+$	25	—	"	$2.82\cdot10^{-11}$	10.55	$2.82\cdot10^{-11}$	10.55	[1]	—
$Co\ (Oxin)_2$	25	—	"	$7.8\cdot10^{-10}$	9.11	$2.2\cdot10^{-20}$	19.66	[1]	—
$Cu\ Oxin^+$	25	—	"	$3.24\cdot10^{-14}$	13.49	$3.24\cdot10^{-14}$	13.49	[1]	[2]
$Cu\ (Oxin)_2$	25	—	"	$1.86\cdot10^{-13}$	12.73	$6.0\cdot10^{-27}$	26.22	[1]	[2]
$La\ Oxin^{3+}$	25	—	"	$2.19\cdot10^{-9}$	8.66	$2.19\cdot10^{-9}$	8.66	[1]	—
$La\ (Oxin)_2^+$	25	—	"	$1.82\cdot10^{-8}$	7.74	$4.0\cdot10^{-17}$	16.40	[1]	—
$Mg\ (Oxin)^+$	25	—	"	$4.17\cdot10^{-7}$	6.38	$4.17\cdot10^{-17}$	6.38	[1]	[2]
$Mg\ (Oxin)_2$	25	—	"	$3.72\cdot10^{-6}$	5.43	$1.55\cdot10^{-12}$	11.81	[1]	[2]
$Mn\ Oxin^+$	25	—	"	$5.25\cdot10^{-9}$	8.28	$5.25\cdot10^{-9}$	8.28	[1]	—
$Mn\ (Oxin)_2$	25	—	"	$6.8\cdot10^{-8}$	7.17	$3.57\cdot10^{-16}$	15.45	[1]	[2]
$Ni\ Oxin^+$	25	—	"	$3.6\cdot10^{-12}$	11.44	$3.6\cdot10^{-12}$	11.44	[1]	—

Complexes with 8-Hydroxyquinoline* (contd.)

Complex ion	Temp. °C	Ionic strength	Method	k	pk	K	pK	References principal	References supplementary
Ni (Oxin)₂	25	—	pH-potent.	$1.15 \cdot 10^{-10}$	9.94	$4.16 \cdot 10^{-22}$	21.38	[1]	[2]
Pb Oxin⁺	25	—	"	$2.46 \cdot 10^{-11}$	10.61	$2.46 \cdot 10^{-11}$	10.61	[1]	—
Pb (Oxin)₂	25	—	"	$8.1 \cdot 10^{-9}$	8.09	$2.0 \cdot 10^{-19}$	18.70	[1]	[2]
Zn Oxin⁺	25	—	"	$1.1 \cdot 10^{-10}$	9.96	$1.1 \cdot 10^{-10}$	9.96	[1]	[2]
Zn (Oxin)₂	25	—	"	$1.26 \cdot 10^{-9}$	8.90	$1.38 \cdot 10^{-19}$	18.86	[1]	[2]
UO₂ Oxin⁺	20	0.3	"	$5.62 \cdot 10^{-12}$	11.25	$5.62 \cdot 10^{-12}$	11.25	[2]	—
UO₂ (Oxin)₂	20	0.3	"	$2.29 \cdot 10^{-10}$	9.64	$1.29 \cdot 10^{-21}$	20.89	[2]	—

* In 50% dioxan-water.

REFERENCES

1. W. D. JOHNSTON and H. FREISER, J. Am. Chem. Soc., 74, 5239 (1952).

2. H. IRVING and H. ROSSOTTI, J. Chem. Soc., 2910 (1954).

T A B L E S

OF

COMPLETE THERMODYNAMIC CHARACTERISTICS

OF

COMPLEX-FORMATION REACTIONS IN SOLUTION

INTRODUCTORY NOTE TO TABLES

All the data given in the Tables relate to reactions of the type:

$$M_{aq} + nA_{aq} = MA_{naq},$$

where M is a metal ion,
A is a ligand, and
n is the coordination number.

All the data are on the basis of a temperature of 25° or room temperature.

The data relating to the stepwise formation of complexes may be obtained by simple arithmetical operations from the figures given, and are therefore not included in the Tables.

In cases where the experimental conditions differ from the foregoing, this fact is indicated in brackets.

The Tables give enthalpy change data only for those cases where the determination has been carried out calorimetrically. Complex compounds from which no data are available for the free energy and entropy changes are not included in the Tables.

The data in the Tables are arranged in groups of complexes having the same ligands, and these groups are arranged in the following order:

COMPLEXES

1. Ammonia
2. Bromide
3. Chloride
4. Cyanide
5. Ethylenediamine
6. Ethylenediaminetetraacetate
7. Fluoride
8. Iodide
9. Oxalate
10. Pyrophosphate
11. Thiosulphate
12. Thiourea
13. Trimethylenediamine

Within each group the metals are arranged in the alphabetical order of their chemical symbols.

The first columns gives the formulae of the complex compounds. In those cases where the ligand formula is cumbersome, arbitrary abbreviations are used, and these are indicated at the start of each group.

The second columns gives the enthalpy change (ΔH) for the reaction indicated above (in calories).

The third column gives the free energy change (ΔG) in calories, calculated from the formula

$$\Delta G = RT \ln K_{instab.},$$

where $K_{instab.}$ is the overall instability constant of the complex in question.

The fourth column gives the entropy change for the same reaction (ΔS) in cal/degree.

The last column indicates the literature references, which are given after the Tables.

Complete Thermodynamic Characterization of
Complex-Formation Reactions in Solution

Complex ion	ΔH, cal.	ΔG, cal.	ΔS, cal/deg.	References
Ammonia Complexes				
$CdNH_3^{2+}$	— 3500	— 3640	0.5	[1]
$Cd (NH_3)_2^{2+}$	— 7000	— 6520	— 1.6	[1]
$Cd (NH_3)_3^{2+}$	—10500	— 8500	— 6.7	[1]
$Cd (NH_3)_4^{2+}$	—14000	— 9780	—14.1	[1]
$Cd (NH_3)_5^{2+}$	—17500	— 9340	—27.1	[1]
$Cd (NH_3)_6^{2+}$	—21000	— 7060	—46.5	[1]
$CuNH_3^{2+}$	— 5600	— 5700	0.3	[2]
$Cu (NH_3)_2^{2+}$	—11100	—10500	— 2.0	[2]
$Cu (NH_3)_3^{2+}$	—16700	— 14480	— 7.4	[2]
$Cu (NH_3)_4^{2+}$	—22000	—17400	—15.3	[2]
$Cu (NH_3)_5^{2+}$	— 27100	—16700	—34.7	[2]
$Hg (NH_3)_2^{2+}$	—24700	—24000	— 2.3	[1]
$Hg (NH_3)_3^{2+}$	—28000	—25400	— 8.7	[1]
$Hg (NH_3)_4^{2+}$	—31600	—26500	—17.0	[1]
$Ni NH_3^{2+}$	— 4000	— 3840	— 0.5	[2]
$Ni (NH_3)_2^{2+}$	— 8000	— 6920	— 3.6	[2]
$Ni (NH_3)_3^{2+}$	—12000	— 9300	— 9.0	[2]
$Ni (NH_3)_4^{2+}$	—16000	—10930	—16.9	[2]
$Ni (NH_3)_5^{2+}$	—20300	—11960	—27.8	[2]
$Ni (NH_3)_6^{2+}$	—24600	—12000	—42.0	[2]
$Zn NH_3^{2+}$	— 2600	— 3260	2.1	[1]
$Zn (NH_3)_2^{2+}$	— 5700	— 6610	3.0	[1]
$Zn (NH_3)_3^{2+}$	—·9600	—10040	1.5	[1]
$Zn (NH_3)_4^{2+}$	—14800	—13000	— 6.0	[1]

Temperature 26.8°C.
Ionic strength μ = 2.0

Bromide Complexes

$HgBr_2$	—22500	—23600	4	[3]
$HgBr_3^-$	—26500	—26800	2	[3]
$HgBr_4^{2-}$	—28700	—28600	0	[3]

Complex ion	ΔH, cal.	ΔG, cal.	ΔS, cal/deg.	References
	Chloride Complexes			
$HgCl_2$	-12300	-17400	17	[3]
$HgCl_3^-$	-13000	-19000	20	[3]
$HgCl_4^{2-}$	-13500	-20300	23	[3]
	Cyanide Complexes			
$Hg\,(CN)_4^{2-}$	-59500	-56300	-10	[3]
$Zn\,(CN)_4^{2-}$	-24700	-22700	-7	[3]
	Ethylenediamine Complexes $\left(H_2NCH_2CH_2NH_2 = En\right)$			
$CuEn^{2+}$	-13030	-14350	4	[11]
$CuEn_2^{2+}$	-25420	-26660	-4	[11]
$NiEn^{2+}$	-9010	-10200	4	[11]
$NiEn_2^{2+}$	-18190	-18800	2	[11]
$NiEn_3^{2+}$	-27190	-24800	-8	[11]
	Ethylenediaminetetraacetate Complexes $([CH_2N(CH_2COO)_2]_2^{4-} = Edta^{4-})$			
$BaEdta^{2-}$	-5100	-10540	18	[13]
$CaEdta^{2-}$	-5800	-14970	31	[13]
$CdEdta^{2-}$	-9100	-20500	38	[13]
$CoEdta^{2-}$	-4100	-21400	58	[13]
$CuEdta^{2-}$	-8200	-24400	55	[13]
$LiEdta^{3-}$	100	-3800	13	[13]
$MgEdta^{2-}$	3100	-12400	52	[13]
$MnEdta^{2-}$	-5200	-17200	41	[13]
$NaEdta^{3-}$	-1400	-2300	3	[13]
$NiEdta^{2-}$	-7600	-24000	55	[13]
$PbEdta^{2-}$	-13100	-23600	35	[13]
$SrEdta^{2-}$	-4200	-11900	26	[13]
$ZnEdta^{2-}$	-4500	-20900	55	[13]
	Fluoride Complexes			
AlF^{2+}	1150	-8370	32	[12]
AlF_2^+	1930	-15220	58	[12]
AlF_3	2120	-20470	76	[12]
AlF_4	2400	-24210	89	[12]
AlF_5^{2-}	1650	-26430	94	[12]
AlF_6^{3-}	100	-27070	91	[12]

Complex ion	ΔH, cal.	ΔG, cal.	ΔS, cal/deg.	References
	Iodide Complexes			
CdJ^+	$-\ 1350$	$-\ 4200$	9	[4]
CdJ_4^{2-}	-10800	$-\ 8400$	$-\ 8$	[5]
HgJ^+	-16600	-18400	6	[4]
HgJ_4^{2-}	-43500	-41300	$-\ 7$	[3]
PbJ^+	$-\ 1000$	$-\ 3130$	7	[4]
PbJ_4^{2-}	-15600	$-\ 8500$	-23	[5]
	Oxalate Complexes $(C_2O_4^{2-}=Ox^{2-})$			
$CoOx_2^{2-}$	$-\ 800$	$-\ 9700$	30	[9]
$CoOx_3^{4-}$	$-\ 3100$	-10800	26	[9]
$CuOx_2^{2-}$	$-\ 1470$	-10900	32	[9]
$MgOx_2^{2-}$	170	$-\ 6000$	21	[9]
$MnOx_2^{2-}$	$-\ 190$	$-\ 7900$	25	[9]
$NiOx_2^{2-}$	$-\ 1100$	-10400	31	[9
$ZnOx_2^{2-}$	$-\ 1300$	-10300	30	[9]
	Pyrophosphate Complexes			
$Cu\ (P_2O_7)_2^{6-}$	$-\ 690$	-12300	39.0	[7]
$Mg\ P_2O_7^{2-}$	2920	$-\ 6280$	31.1	[8]
$Ni\ P_2O_7^{2-}$	4220	$-\ 7950$	40.8	[7]
$Ni\ (P_2O_7)_2^{6-}$	2000	$-\ 9800$	39.6	[7]
$Pb\ (P_2O_7)_2^{6-}$	$-\ 1030$	$-\ 7250$	20.9	[7]
$Zn\ (P_2O_7)_2^{6-}$	2640	$-\ 8860$	38.6	[7]
	Thiosulphate Complexes			
CdS_2O_3	000	$-\ 3700$	12	[10]
$Cd\ (S_2O_3)_2^{2-}$	$-\ 1500$	$-\ 7090$	18	[10]
$Cd\ (S_2O_3)_3^{4-}$	$-\ 3400$	$-\ 8560$	17	[10]
ZnS_2O_3	$-\ 3500$	$-\ 1690$	17	[10]

Complex ion	ΔH, cal.	ΔG, cal.	ΔS, cal/deg.	References
Thiourea Complexes				
$Ag(CSN_2H_4)_3^+$	−30700	−17900	−43	[9]
$Bi(CSN_2H_4)_6^{3+}$	−22000	−16200	−19.5	[9]
$Cd(CSN_2H_4)_3^{2+}$	−13000	− 4000	−30	[9]
$Cu(CSN_2H_4)_3^+$	−32300	−17500	−50	[9]
$Hg(CSN_2H_4)_4^{2+}$	−50000	−38000	−40	[9, 4]
$Pb(CSN_2H_4)_3^{2+}$	−13300	− 2400	−36	[9]
Trimethylenediamine Complexes ($H_2NCH_2CH_2CH_2NH_2$ = Tmen)				
$CuTmen_2^{2+}$	−22760	−23400	2	[11]
$NiTmen^{2+}$	− 7770	− 8700	3	[11]
$Ni(Tmen)_2^{2+}$	−15000	−14700	− 1	[11]
$Ni(Tmen)_3^{2+}$	−21340	−16400	−17	[11]

REFERENCES

1. K.B.YATSIMIRSKII and P.M.MILYUKOV, Zh.neorg.khim., 2, 1046 (1957).
2. K.B.YATSIMIRSKII and P.M.MILYUKOV, Zh.fiz.khim., 31, 842 (1957).
3. M. BERTHELOT, Thermochimie, Paris (1897).
4. K.B.YATSIMIRSKII and A.A.SHUTOV, Zh.fiz.khim., 28, 30 (1954).
5. K.B.YATSIMIRSKII and A.A.ASTASHEVA, ibid., 26, 239 (1952).
6. E.K.ZOLOTAREV, The Study of Oxalate Complexes in Solution (Issledovaniye oksalatnykh kompleksov v rastvore), Thesis, Institute of Chemical Technology, Ivanovo (1955).
7. K.B.YATSIMIRSKII and V.P.VASIL'EV, Zh.fiz.khim., 30, 901 (1956).
8. V.P.VASIL'EV, ibid., 31, 692 (1957).
9. K.B.YATSIMIRSKII and A.A.ASTASHEVA, ibid., 27, 1539 (1953).
10. K.B.YATSIMIRSKII and L.V.GUS'KOVA, Zh.neorg.khim., 2, 2039 (1957).
11. J.POULSEN and J.BJERRUM, Acta Chem. Scand., 9, 1407 (1955).
12. W.M.LATIMER and W.L.JOLLY, J.Am.Chem.Soc., 75, 1548 (1954).
13. R.G.CHARLES, ibid., 76, 5854 (1954).

INDEX OF TABULATED LIGANDS